LAIRDS, LAND
AND SUSTAINABILITY

SCOTTISH PERSPECTIVES ON UPLAND MANAGEMENT

Edited by Jayne Glass, Martin F. Price,
Charles Warren and Alister Scott

EDINBURGH
University Press

© editorial matter and organisation
Jayne Glass, Martin F. Price, Charles Warren
and Alister Scott, 2013
© the chapters their several authors, 2013

Edinburgh University Press Ltd
22 George Square, Edinburgh EH8 9LF
www.euppublishing.com

Typeset in 11/13 Minion Pro by
Servis Filmsetting Ltd, Stockport, Cheshire
and printed and bound in Great Britain by
CPI Group (UK) Ltd, Croydon CR0 4YY

A CIP record for this book is available from the
British Library

ISBN 978 0 7486 4591 6 (hardback)
ISBN 978 0 7486 4590 9 (paperback)
ISBN 978 0 7486 8588 2 (webready PDF)
ISBN 978 0 7486 8589 9 (epub)
ISBN 978 0 7486 8590 5 (Amazon ebook)

The right of the contributors
to be identified as authors of this work
has been asserted in accordance with
the Copyright, Designs and Patents Act 1988.

Contents

Expanded contents list

Preface

This book represents the final synthesis of 'Sustainable Estates for the 21st Century', the largest in-house project undertaken to date at the Centre for Mountain Studies (CMS) at Perth College, University of the Highlands and Islands (UHI). Having established the CMS in 2000, I believed that it was important that its staff and students should be involved in a major project on a theme that was specific to Scotland's mountains. After considerable reading and discussion, I realised that a project on large upland estates would be most appropriate. The first reason is that Scotland has the most concentrated pattern of private landownership in the world, a pattern that is even more exaggerated in the mountains, especially the Highlands. Over recent decades, some of these estates have been purchased by conservation non-governmental organisations (NGOs) and local communities, and the largest single landowner – though only partly in the mountains – is Forestry Commission Scotland. Nevertheless, private ownership remains dominant, a phenomenon that particularly characterises Scotland's mountains. The second reason is that the ownership, and the related management, of land are key issues in Scotland, with a long and complex history tied to national identity. Consequently, after devolution in 1999, one of the priorities of the renewed Scottish Parliament was the Land Reform (Scotland) Act 2003. Until the present project, no substantial study of estates had been undertaken since its passage. Finally, though there have been many studies of individual estates, or specific aspects of estate management, and some studies based on samples of tens or hundreds of landowners, there had never been a project that looked at a considerable number of estates – privately, community, and NGO owned – using common methodologies and aiming to draw overall conclusions about the implications of different types of landownership for the sustainable management of estates. Again, given the passage of the Land Reform (Scotland) Act 2003, such an integrated project appeared particularly timely.

The implementation of the project was made possible by the remarkable generosity of Henry Angest, the owner of an estate in Highland Perthshire. In late 2006, following discussions with Mandy Exley, then Principal of Perth College, UHI, and myself, Mr Angest agreed to fund fully a project including four studies – two on privately owned estates, and one each on community- and NGO-owned estates – and a one-year synthesis phase. Each of the students who would undertake the studies would be allowed to choose their own topic for a PhD. We were fortunate to recruit four excellent students though, ultimately, one was not able to complete the PhD and the work she had started on community-owned estates was taken on by Rob Mc Morran who had previously completed his PhD with the CMS. The two PhD students considering privately owned estates first undertook a major survey of their owners, facilitated by the Scottish Rural Property and Business Association (SRPBA), and then addressed the motivations of eleven landowners (Pippa Wagstaff) and the potential for positive interactions between six estates and the local communities (Annie McKee). The third PhD student, Jayne Glass, developed a sustainability tool for the owners and managers of upland estates, working with nineteen experts from across Scotland, and tested this on two NGO-owned estates. The four projects were co-ordinated and supervised by myself, with substantial inputs from Charles Warren (University of St Andrews) and Alister Scott (Birmingham City University), as PhD co-supervisors. To ensure the on-the-ground and policy relevance of the project, an advisory board was established, with representatives from the SRPBA (now Scottish Land and Estates, SLE: Richard Cooke), Scottish Environment LINK (Bob Aitken), Cairngorms National Park Authority (Hamish Trench), the Knoydart Trust (Angela Williams), and the Scottish Government (Gerry Selkirk). The members of the board were involved in selecting the students, and providing feedback and advice throughout the project. I would like to express many thanks: to Henry Angest, Mandy Exley, the members of the advisory board, my co-supervisors, Douglas McAdam and other staff at SLE, and the hundreds of people who gave their time to contribute information, evidence and advice to the four primary researchers.

In mid 2010, as the PhD theses were being written up, Annie McKee decided to submit a proposal to the Economic and Social Research Council (ESRC) for a Knowledge Exchange Small Grant to ensure that, on the one hand, the outcomes of her research would be fed back to the communities where she had worked and, on the other, that good practices in collaboration between estates, communities and other partners could be discussed and compiled in a 'user-friendly' publication. Wider interest for this additional activity was shown by the fact that the Scottish Government and SLE both agreed to provide match funding. The application was successful, leading to three workshops in 2011 and, in 2012, a booklet that has been very widely distributed in Scotland and beyond. Again, I would like to thank the ESRC, the Scottish Government, and SLE for their support for this work.

Finally, I would like to thank Annie McKee, Rob Mc Morran, and Pippa Wagstaff for all their enthusiasm and hard work over the past five years, and especially Jayne Glass for taking on the further challenging role of compiling and acting as lead editor of this book. Published ten years after the Land Reform (Scotland) Act 2003, I believe it provides a significant contribution to the understanding of the complex challenges faced by those owning, managing, living on and near, and visiting Scotland's upland estates. Inevitably, it raises many issues for research and analysis to be undertaken in future projects. Equally, though the focus of the book is specifically Scottish, I believe that many of the issues it addresses are relevant for other rural and mountain areas elsewhere in Europe and more widely.

Martin F. Price

Acronyms

CAP	Common Agricultural Policy
COAT	Cairngorms Outdoor Access Trust
CSR	Corporate Social Responsibility
DCS	Deer Commission Scotland (now part of Scottish Natural Heritage)
DMG	Deer Management Group
EEA	European Environment Agency
EU	European Union
FCS	Forestry Commission Scotland
FTE	Full-time equivalent
GVA	Gross value added
HEP	Hydroelectric power
JMT	John Muir Trust
LFA	Less Favoured Areas
LRPG	Land Reform Policy Group
LRRG	Land Reform Review Group
LRSA	Land Reform (Scotland) Act 2003
MEA	Millennium Ecosystem Assessment
MFFP	Moors for the Future Partnership
NEA	National Ecosystem Assessment
NGO	Non-governmental organisation
NSA	National Scenic Area
NTS	National Trust for Scotland
RSPB	Royal Society for the Protection of Birds
SAC	Special Area of Conservation
SLE	Scottish Land and Estates
SNH	Scottish Natural Heritage
SOAC	Scottish Outdoor Access Code
SPA	Special Protection Area
SRDP	Scotland Rural Development Programme
SWT	Scottish Wildlife Trust
UNCED	United Nations Conference on Environment and Development
UNESCO	United Nations Educational, Scientific, and Cultural Organization
VAT	Value Added Tax

Tables

Figures

Boxes

Notes on the contributors

Dr Jayne H. Glass
Jayne Glass is a Research Associate at the Centre for Mountain Studies, Perth College, University of the Highlands and Islands, Scotland, and holds degrees in geography and in environmental sustainability from the universities of Oxford and Edinburgh. She recently completed a PhD in Sustainability Studies at the University of the Highlands of Islands on knowledge co-production for sustainable upland estate management in Scotland (awarded by the University of Aberdeen).

Professor Martin F. Price
Martin Price is Director of the Centre for Mountain Studies, Perth College, University of the Highlands and Islands, Scotland, and holds the UNESCO Chair for Sustainable Mountain Development. He previously worked at the universities of Oxford, Bern, and Colorado, and at the National Center for Atmospheric Research. Books he has edited include *Mountain Geography: Physical and Human Dimensions* (2013), *Mountain Area Research and Management* (2007), *The Mountains of Northern Europe* (2005), and *Key Issues for Mountain Areas* (2004).

Dr Charles R. Warren
Charles Warren is Senior Lecturer in the Department of Geography and Sustainable Development at the University of St Andrews, and holds degrees in geography, glaciology, and resource management from Oxford and Edinburgh universities. He has written widely on Scottish land-use issues, including his book *Managing Scotland's Environment* (Edinburgh University Press, 2009). He also co-edited *Learning from Wind Power: Governance, Societal and Policy Perspectives on Sustainable Energy* (2012).

Professor Alister J. Scott
Alister Scott is Professor of Environment and Spatial Planning at Birmingham City University. He is a chartered planner with roots firmly in geography. His research and teaching are located within an interdisciplinary framework with a focus on complex and messy policy and land-use problems. He has become an expert at the interface of spatial planning and the ecosystem approach with projects exploring the rural urban fringe as part of the RELU programme (2009–11) and the embedding of the ecosystem approach in tools for improved policy- and decision-making as part of the National

Ecosystem Assessment (NEA) Follow-on project 2012–14. He also sits on the NEA expert panel and is a communication adviser for the NERC BESS programme.

Annie J. McKee

Annie McKee is an early career researcher and Research Assistant in the Social, Economic and Geographical Science Group at the James Hutton Institute in Aberdeen, Scotland. She holds degrees in geography and sustainable rural development from St Andrews and Aberdeen universities, and is currently completing her PhD with the Centre for Mountain Studies at Perth College, University of the Highlands and Islands. Annie is a Trustee of the Andrew Raven Trust, and board member of the Rural Housing Service.

Pippa K. Wagstaff

Pippa Wagstaff is a PhD researcher at the Centre for Mountain Studies, Perth College, University of the Highlands and Islands, Scotland and holds a first degree from the Open University and a Masters in Land Economy from the University of Aberdeen. Pippa had a previous career as a chartered accountant. She is currently completing her PhD on landowner motivation as part of the Sustainable Estates project.

Dr Robert Mc Morran

Rob Mc Morran is a Research Associate at the Centre for Mountain Studies, Perth College, University of the Highlands and Islands. Rob is an interdisciplinary researcher with experience in qualitative and GIS methodologies and specific interests in conceptions of wildness; sustainable rural governance and multifunctional land use; and policy processes relating to rural and mountainous regions. Rob has acted as a consultant to Scottish Natural Heritage, Forestry Commission Scotland, Scotland's National Park Authorities, the Scottish Countryside Alliance, and the John Muir Trust. Rob's research is predominantly applied, with recent research on mountain food-supply chains in Europe resulting directly in changes to EU Policy. Rob is an Honorary Research Associate of the Wildland Research Institute (University of Leeds) and a member of the BBC Scotland Rural Affairs Committee.

PART ONE
Sustainability in the uplands

CHAPTER ONE

Sustainability in the uplands: introducing key concepts

Jayne Glass, Alister Scott, Martin F. Price and Charles Warren

INTRODUCTION

Uplands are special and valued places. Whether we live in or near them, we depend on them in many different ways to meet our daily needs. Uplands are multifunctional landscapes, providing resources and livelihoods to local people, as well as vital eco- system services[1] to society, such as clean water, hydroelectricity, flood control, carbon sequestration, biodiversity, recreation, and aesthetic and cultural services (Körner et al. 2005; EEA 2010; Schild and Scharma 2011). Recent research in northern Europe – particularly in the fields of climate change, biodiversity and soil management – has illuminated the importance of upland areas, in terms of both their vulnerability to environmental change and their major influences on lowland ecologies and econo- mies (e.g. Burt 2001; Carling et al. 2001; Werritty 2002; McVittie et al. 2005; Maxwell and Birnie 2005; Reed et al. 2009). They are predominantly open and extensively managed landscapes, and they are places of high scenic and environmental quality, recognised by numerous international and national conservation designations. For these reasons, and also because of their 'wild' appeal, uplands are also important set- tings for recreation which supports many livelihoods in local communities.

Numerous global driving forces have impacts on upland areas, particularly in the context of increasing pressures to meet growing demands for food, timber and recrea- tion (Bonn et al. 2009a; Reed et al. 2009). Globally, over the past two decades, many upland regions appear to have developed to a lesser extent than adjacent lowland regions, partly as a result of their protected status, along with the complex interplay of different historical, geopolitical, economic, environmental and sociocultural factors (Maselli 2012). Threats related to climate change, habitat degradation and loss of public funding have implications for upland habitats, economies and communities, and land-management decisions are critical in determining the balance of activities that take place in upland regions. Those who own land and have control over its man- agement have considerable influence and responsibility regarding land-use decisions, set within planning and environmental policy frameworks. Given the large number of ecosystem services linked to upland regions, however, there are contested views about land use among many different stakeholders (Bonn et al. 2009b). Property rights are therefore of fundamental importance as they have an impact on the status of land as the 'ultimate resource' from which all prospects for development, production and conservation in uplands are derived.

This chapter sets the context for the whole volume, providing an introduction to principles and practices related to the sustainable governance of upland areas. Set within the dynamics of significant environmental, social, economic and political change, the focus is on how sustainability has been conceptualised and operationalised. 'Sustainability' is increasingly interpreted as a prerequisite for delivering the wide range of ecosystem services that are linked to upland areas (Maxwell and Birnie 2005; Reed et al. 2011) (Box 1.1). Thus, the importance of involving multiple stakeholders in the governance of these areas, particularly through collaborative, landscape-scale approaches, is favoured and critically discussed. In addition, the importance of land-owners' decisions and actions is considered, leading to a discussion of property rights and responsibilities within a range of landownership models in upland Scotland.

THE FOCUS OF THIS BOOK

This book focuses on upland regions in Scotland, and pursues the general proposition that improved policy-making and decision-making are needed to facilitate sustainable management and governance of upland areas. The content of the chapters is largely based on the results of the four studies comprising the 'Sustainable Estates for the 21st Century' research project, carried out at the Centre for Mountain Studies at Perth College, University of the Highlands and Islands, from 2007 to 2012. Much of the land in the Scottish uplands is held in privately owned estates, some of which are many tens of thousands of hectares in area. Estates are also owned by conservation organisations, public bodies and, increasingly since the Land Reform (Scotland) Act 2003, local communities (more detail is given at the end of this chapter). Thus, recognising that Scotland's uplands have a diverse pattern of landownership, the aim of the project was to investigate the alignment of the management of large, upland 'estates' with the concept of sustainability, in order to understand how best to manage the relationships between people, place and economy in this context. The research explores a number of questions related to the governance of upland areas and discusses the implications of this pattern of landownership and the resulting impacts of landowner/manager decisions on upland environments, economies and communities.

This necessitates improved understandings of upland systems and of how people, place and economy interact within a changing and increasingly complex governance agenda. Here, issues of power, equity and inclusivity come to the fore (Schild and Scharma 2011). The book explores critically the human-environment dimensions and interactions of upland governance but detailed attention is not given to ecological processes in upland regions. Those seeking a more thorough understanding of physical driving forces of change in upland regions in northern Europe, including Britain, should consult Thompson et al. (2005), Bonn et al. (2009), EEA (2010), Marrs et al. (2011) and Mansfield (2011).

Box 1.1 Ecosystem services in uplands (Adapted from:
Bonn et al. 2009a)

The concept of ecosystem services has been developed to 'aid the understanding of
the human use and management of natural resources' (Bonn et al. 2009a: 2). The
Millennium Ecosystem Assessment (2005) identified four main types of ecosystem
services that directly affect human well-being: supporting services; provisioning
services; regulating services; and cultural services. Figure 1.1 shows these services
in an upland context.

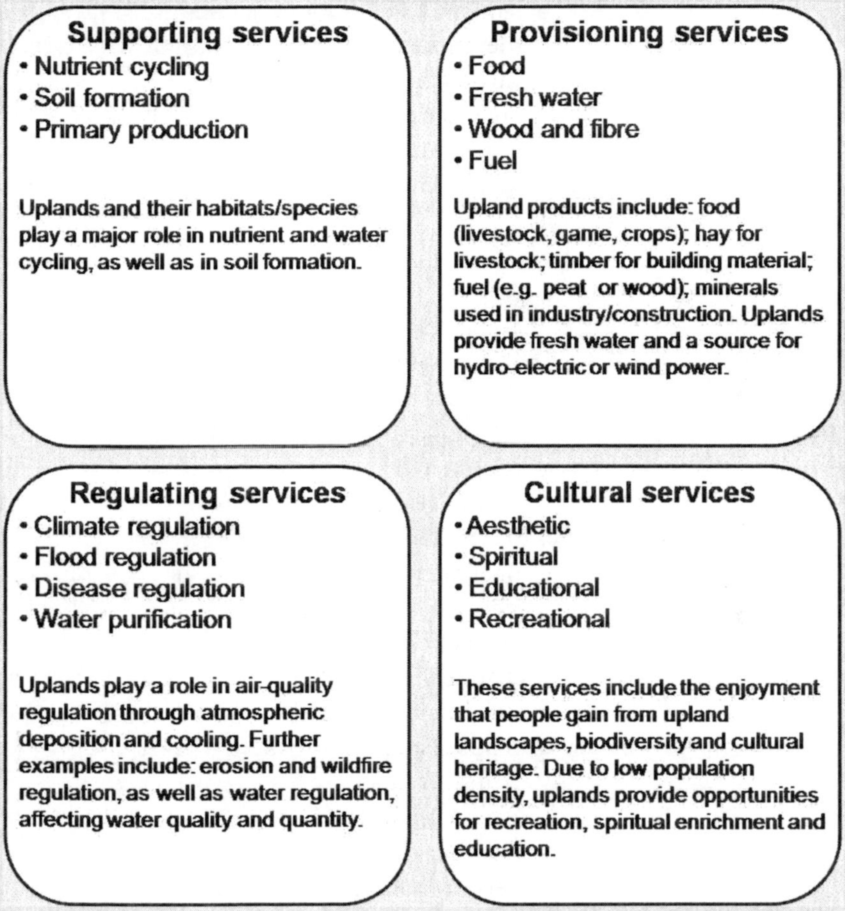

Supporting services
- Nutrient cycling
- Soil formation
- Primary production

Uplands and their habitats/species
play a major role in nutrient and water
cycling, as well as in soil formation.

Provisioning services
- Food
- Fresh water
- Wood and fibre
- Fuel

Upland products include: food
(livestock, game, crops); hay for
livestock; timber for building material;
fuel (e.g. peat or wood); minerals
used in industry/construction. Uplands
provide fresh water and a source for
hydro-electric or wind power.

Regulating services
- Climate regulation
- Flood regulation
- Disease regulation
- Water purification

Uplands play a role in air-quality
regulation through atmospheric
deposition and cooling. Further
examples include: erosion and wildfire
regulation, as well as water regulation,
affecting water quality and quantity.

Cultural services
- Aesthetic
- Spiritual
- Educational
- Recreational

These services include the enjoyment
that people gain from upland
landscapes, biodiversity and cultural
heritage. Due to low population
density, uplands provide opportunities
for recreation, spiritual enrichment and
education.

FIGURE 1.1 *Ecosystem services in uplands*

All these services are linked to human well-being because people depend on them.
The Millennium Ecosystem Assessment noted that the four types of services are
linked to: security; basic material for good life; health; good social relations; and
freedom of choice and action. The full benefits that people gain from these services
are difficult to measure, but unsustainable management practices risk the loss of
these services for current and future generations (Bonn et al. 2009a).

The first section of the book presents and discusses the wider context of the research, focusing particularly on the governance of upland regions, exploring critically the concept of sustainability, both globally and in Britain (Chapter 1), and identifying key issues for the sustainable management of Scotland's uplands in particular (Chapter 2). The second section focuses on private landownership in upland Scotland, providing insight into the contemporary Scottish private estate (Chapter 3), the motivations of private landowners (Chapter 4), and the roles that private landowners can play in encouraging sustainable communities in upland regions (Chapter 5). The third section turns towards 'social ownership' of land by local communities (Chapter 6) and conservation organisations (Chapter 7), exploring the current and potential contributions of these newer models of landownership. Finally, in the fourth section, a sustainability tool for the owners/managers of upland estates is proposed (Chapter 8), and the key outcomes of the whole study are discussed (Chapter 9).

Defining uplands

Defining what we mean by 'uplands' is not an easy task, as many different definitions and classifications apply to upland areas (Mansfield 2011). If one starts with their geology, most British uplands are found on rocks that are resistant to erosion (Bunce et al. 1996). In other European languages, uplands would be described as *moyenne montagne* in French or *Mittelgebirge* in German. Both of these can be translated as 'middle mountains' which tend to have a rounded relief, primarily resulting from glacial erosion in the last ice age. Other examples in Europe include the Vosges, the Black Forest, the Giant Mountains, and the Massif Central. They can be contrasted with 'high mountains' (*haute montagne, Hochgebirge*), such as the Alps and other high ranges up to those of the Himalaya, with much more jagged and rugged landscapes where glaciation is often still under way (Ives et al. 1997). Essentially, uplands are mountains and, as noted in a report published in 2010 by the European Environment Agency (EEA), while there is widespread agreement that the summits of high mountains are, indeed, mountains, there are 'contrasting opinions regarding both the difference between mountains and hills and, particularly, the lower extent of these topographical features of the landscape' (p. 26). For the EEA report, following previous work at global and European scales, mountains were defined according to different thresholds for altitude, climate and topography. The resulting map of European mountain areas is shown in Figure 1.2.

European mountains have also been defined in legal documents, though not in the United Kingdom. In particular, since 1975, the European Commission has published a number of definitions of 'mountains', first in Council Directive 75/268/EEC on 'mountain and hill farming and farming in certain less favoured areas' and, most recently, in 2005:

Mountain areas shall be those characterised by a considerable limitation of the possibilities for using the land and an appreciable increase in the cost of working

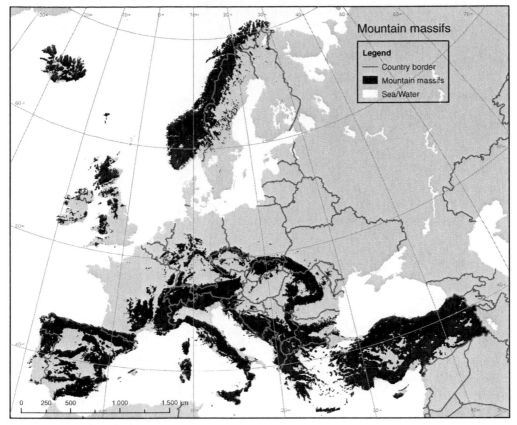

FIGURE 1.2 *European mountain massifs (Source: European Environment Agency 2010 – Owner: Environment Agency Austria)*

it, due to: (a) the existence, because of the altitude, of very difficult climatic conditions, the effect of which is substantially to shorten the growing season; (b) at a lower altitude, the presence over the greater part of the area in question of slopes too steep for the use of machinery or requiring the use of very expensive special equipment, or a combination of these two factors, where the handicap resulting from each taken separately is less acute, but the combination of the two gives rise to an equivalent handicap. (Article 50, Council Regulation 1698/2005 on support for rural development by the European Agricultural Fund for Rural Development)

This designation of mountains as Less Favoured Areas (LFAs) describes areas where farming becomes marginal and less profitable because the productivity of the land is limited, often severely, by physical factors such as harsh climate, short growing season, poor soil fertility and drainage, steep slopes, and high altitudes. Using this designation, a third of the land surface of Britain can be deemed 'upland' in character (Reed et al. 2009). A map of LFAs in Scotland is shown in Figure 1.3.

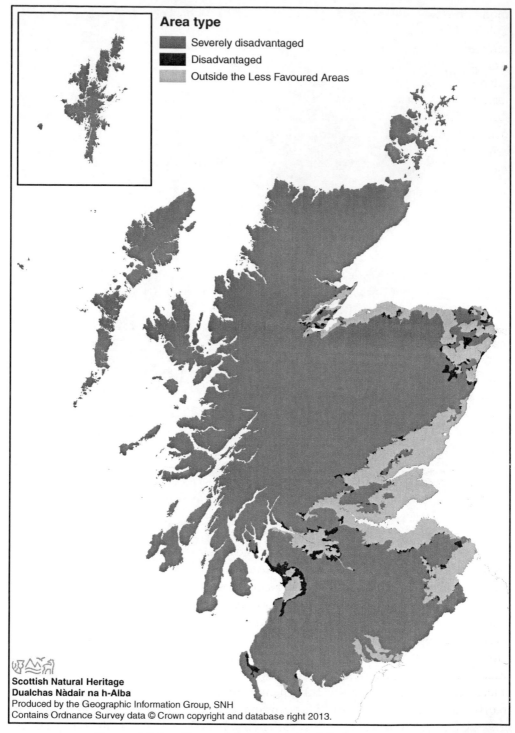

FIGURE 1.3 *Less Favoured Areas in Scotland (Source: SNH 2013)*

Though this is, above all, a socio-economic designation, it corresponds well with an ecological approach to defining the British uplands: upland habitats, as defined by Ratcliffe (1977) and Ratcliffe and Thompson (1988), are areas above the upper limits of enclosed farmland. According to this ecological definition, about half of Scotland is upland in character, comprising a mix of blanket bog, rough grassland, dwarf-shrub heath and a range of other mountain and moorland habitats (Marrs et al. 2011). For our work in Scotland, we used a definition related primarily to land use: the research focused on upland estates identified by combining location above the upper limits of farmland with elevation above sea level (predominantly above 185 metres) and the boundary of the LFA designation.

HUMAN DIMENSIONS OF CHANGE IN UPLAND REGIONS

Upland landscapes are the product of dynamic interactions between natural and human influences, affected by a wide range of international, national and local management policies.[2] In 1988, Ratcliffe and Thompson identified six main elements of these interactions in the British uplands: extensive forest clearance; extensive use of grazing range by domestic livestock; land improvement for agriculture and hunting/ field sports; persecution of wildlife (especially predators) relating to livestock and game management; industrial acidification; and extensive conifer afforestation. Today, we can add an increase in tourism and related recreation, restoration of semi-natural habitats, and the expansion of renewable energy developments, notably hydroelectric schemes and wind farms (Thompson et al. 2005). These are all examples of direct interactions of people with the uplands; more indirect interactions are through air pollution, particularly nitrogen deposition, and anthropogenic climate change (Brooker 2011).

Upland regions provide a microcosm of the ways in which European societies view the countryside and the activities that take place there. In the last few decades, there has been a movement away from a dominant social paradigm that views the countryside as a place for the *production* of food, fibre and timber, towards comparable regard as a place of *consumption* of recreation and amenity, where 'habitats, wildlife, water and landscape have intrinsic worth and value' (Maxwell and Birnie 2005, 21; Midgley et al. 2008). This has also been seen as a shift from 'productivism' to 'post-productivism' (Mather et al. 2006) although, more recently, the pendulum has swung back towards productivism somewhat, not least as a result of rising concerns over food security and 'food miles'. As a result, the challenges in upland areas have become increasingly complex as these two functions intersect in the same geographical space with increased contestation (Scott et al. 2009). Changes in the way the countryside is utilised and perceived have been attributed to: market forces and technology (Mather et al. 2006); government policy and expenditure (Hellman and Verburg 2010); people in urban areas looking more and more to rural areas for recreational purposes (Skuras et al. 2009); and an increasingly vocal environmental lobby (Mowle 1997). In this multifunctional and partly 'post-productivist' landscape, which generally emerged earlier in the uplands than in the agriculturally more productive lowlands, ecosystem

---

Box 1.2 A shift to post-productivist landscapes in Norway

Traditionally, the upper parts of the mountains of Norway were used for grazing livestock. As with similar systems in mountain regions around the world, the animals were moved to higher altitudes over the summer. Over centuries, this pattern of land use led to the loss of birch forests as the heathlands and grasslands used for grazing expanded. Since the mid twentieth century, this land use has largely disappeared due to changes in agricultural practices and a demand for high-yielding cattle which are now mainly fed from arable land. As a result, the mountain landscape has changed significantly: pine and birch forests are taking over the heathlands and grasslands. Sheep are still grazed in the mountains but on the higher alpine tundra (Olsson 2005). Over the same period, there have been considerable increases both in the number of protected areas in these mountains (Eiter and Potthoff 2007) and in tourism and recreation, especially along the extensive trail networks linking the many mountain huts, and in locations with the best road access where extensive areas with second homes have been developed: about six thousand are being built each year (Rye and Berg 2011).

---

services are given increasing attention and consideration (Scottish upland ecosystem services are considered in more detail in Chapter 2).

In a European context, the situation differs from one nation to another, though some recurring themes underpin national approaches to the management of rural areas: agricultural modernisation; infrastructure development; maintaining regional populations; landscape protection and management; rural economic diversification; social and economic cohesion; and relieving rural disadvantage and deprivation (Maxwell and Birnie 2005). Across Europe, the promotion of non-agrarian over agrarian uses of rural land has been a growing trend (Tovey and Mooney 2006), and agriculture is increasingly considered from the point of view of a diversified rural economy shaped by the demands of consumption (Slee 2005). Box 1.2 provides an example from Norway's mountain areas. The general thrust of European Union (EU) policy and of proposed continuing reform of the Common Agricultural Policy centres around the replacement of production subsidies with area-based and environmental payments, with an emphasis on sustainable land use. Agricultural production has therefore lost the pre-eminent position that it once enjoyed, on the basis that agriculture provides benefits and services other than food commodities (Quinn et al. 2010; Hart et al. 2011). Agricultural productivity is still a key concern on good-quality land but agricultural decline or land abandonment has become common on poorer or marginal land. This is particularly apparent in the British uplands where sheep-stocking numbers have declined across many areas (SAC 2008). There has been a general relative decline in the economic importance of agriculture in industrialised, densely populated EU member states such as the United Kingdom, Germany and France (Maxwell and Birnie 2005). In counterpoint, there has been increasing emphasis on environmental protection and conservation (and the implementation of the EU

Birds and Habitats Directives), and on the amenity quality of upland landscapes and the role of local amenity attractions in driving economic change (Curry 2009).

This shift in how we conceptualise the uplands is also reminiscent of a shift from exogenous ('top-down') models of rural development theory and policy towards more neo-endogenous ('bottom-up') development (Ray 2008). Challenges in upland and other rural areas are increasingly being tackled in integrated ways, with public interest extending beyond environmental concerns to include broader social and cultural interests. For example, uplands are often remote, fragile, and physically testing environments that provide a sense of wildness, refuge, solitude and a chance to 'get away from it all'. This has led to an increase in recreation in uplands and growing interest in, and emotional attachment to, these regions: they can function as a 'lightning rod for the contested nature of the arguments that have to do with how rural land is used, who owns it, and who has rights of access to it' (Maxwell and Birnie 2005, 21). The recreational and amenity value of uplands also creates high demand for residential developments (particularly holiday accommodation and 'second homes'). Such development in turn creates additional stresses on water and ecological values as a result of increased road development, traffic, and fragmentation of landscapes. Key to the control and regulation of these demands are the governance systems and the ways that planning shapes the landscape. We consider these below.

There is now wide appreciation of the need to focus on process-led, multi-sectoral and territorial approaches which emphasise subsidiarity (decision-making at the lowest level) and are rooted in local resources and people (Shucksmith 2000; Bryden and Geisler 2007; Perry 2007; Satsangi 2009). This realisation has led to an increasing focus on the concept of environmental governance which recognises the importance of involving a wide range of stakeholders in environmental decision-making (Carter 2006). Warren (2009) illustrates this well: 'Local decisions were once dictated mainly by local considerations, but in recent decades the spatial scale of management perspectives has rapidly expanded' (p. 51).

Upland management exists within a context of diverse private/public preferences and uncertainty about future sources of income and financial support. There are many people who have stakes in how the uplands are managed, and it is therefore increasingly 'necessary to explore the range of emerging approaches to rural governance that align community interests and rural planning' (Reed et al. 2009, S213). While inequalities between stakeholders and conflicting objectives can all work as barriers to effective land management for the public good (Ostrom et al. 1999; Quinn et al. 2007), there is a growing need to develop more effective and joined-up approaches to upland governance which facilitate the sustainable management of these regions.

WHAT DOES A SUSTAINABILITY AGENDA IMPLY FOR THE GOVERNANCE OF UPLAND AREAS?

The concept of sustainability (and hence 'sustainable' management) is difficult to put into practice which may explain why widely accepted, overarching definitions remain elusive, unsatisfactory and contested, leading to the advancement of new ideas, such as resilience to reinvigorate an increasingly sterile academic discourse (Fazey 2010). Nevertheless, the concept of sustainability has provided one of the most unifying global paradigms of recent times, having been described as a 'multi-dimensional bridging concept' (Meadowcroft 2000, 381) or a 'grand compromise' (Kates et al. 2005). It has also been criticised for its inherent anthropocentricism, however, and for being contested, shaped to fit a spectrum of world views, and too generic (Maxwell and Cannell 2000).

There is general agreement that sustainability is a process that enables current needs to be satisfied while maintaining long-term perspectives regarding the use and availability of natural and other resources into the future. The widely cited definition of sustainable development from 'Our Common Future' highlights the importance of equity between the well-being of current and future generations, with a focus on development that 'meets the needs of the present without compromising the ability of future generations to meet their own needs' (World Commission on Environment and Development 1987). In the twenty-six years since the publication of this report, the political rhetoric of the sustainability agenda has become embedded in academic literature and policy development at global, EU and UK scales (Luke 2005). Its meaning and relevance have arguably been lost at the regional and local scales, however, with top-down narratives that stem from national levels perceived as remote from the local context:

> . . . many of the key details of a sustainable lifestyle will be idiosyncratic to a particular community located in a particular place with a distinctive ecology and will be intelligible only as a part of the history of that place, which we have interpreted as including both the landscape and the peoples that live there. (Norton 2005, 358–9)

Applying these principles to the uplands, Price and Kim (1999, 205) suggest that sustainable mountain development is 'a regionally specific process of sustainable development that concerns both mountain regions and populations living downstream or otherwise dependent on these regions in various ways'. More detail about the concept of sustainable mountain development is given in Box 1.3. This region-specific definition is important because it recognises that, while uplands may be distinct geographically, their resources in the sense of 'public goods' can be regarded as 'belonging' to a wider community than those residing or having ownership in upland areas (Maxwell and Birnie 2005). From a sustainability perspective, this implies a need to find sensible and equitable ways of sharing and managing public goods for both present and future generations, and, arguably, to distribute property rights equitably and responsibly within a range of landownership models. Crucially, sustainable mountain development requires the integration of various objectives of land management and the involvement of key stakeholders in decision-making.

Sustainability presents an opportunity for changing the current culture and practice of policy and decision-making that Scott et al. (in press) view as leading to 'disintegrated development'. Upland areas have challenges as well as opportunities related

---

Box 1.3 Sustainable Mountain Development

Mountain regions occupy 24 per cent of the world's land surface and are home to 12 per cent of the world's population (Schild and Scharma 2011). In the EU, these figures are 29 per cent and 13 per cent respectively (EEA 2010). More than half the world's population depends on fresh water that is captured, stored and purified in mountain regions; mountain regions are 'hotspots of biodiversity' (e.g. 43 per cent of all Natura 2000 designated sites across the EU are located in mountains); and are key global destinations for tourist and recreation activities (Grêt-Regamey et al. 2012).

Mountains have now assumed global significance as key components of sustainable development agendas, and environmental, economic and social issues in mountain areas have recently received wide attention under the umbrella concept of 'sustainable mountain development' (Debarbieux and Price 2008). Developed from global debates and a political spotlight on 'sustainability', the concept has evolved in tandem with growing concern about upland ecosystems and biodiversity, upland economies, communities and cultural heritage.

Agenda 21, a 'blueprint' for sustainable development agreed at the United Nations Conference on Environment and Development (UNCED) in Rio de Janeiro in 1992, includes a chapter on 'Managing Fragile Ecosystems: Sustainable Mountain Development' (Chapter 13). This recognises that mountains are fragile ecosystems that matter for humankind (Maselli 2012), signifying, for the first time, that mountains should be 'accorded comparable priority in the global debate about environment and development with issues such as global climate change, desertification and deforestation' (Price and Kohler 2013). The two 'programme areas' of Chapter 13 focus specifically on generating and strengthening knowledge about the ecology and sustainable development of mountain ecosystems, and promoting integrated watershed development and alternative livelihood opportunities.

---

to new discourses of what we want our uplands to be (see Box 1.4 for a summary of the key 'ingredients' that are relevant to the sustainable management of public goods in upland areas). Discourses of 'needs' and 'rights' are central tenets that should be examined, and the concept of Corporate Social Responsibility (CSR) has emerged as a product of the social justice debate to harness the potential positive contribution of upland businesses in a manner that deters corporate irresponsibility (Newell and Frynas 2007). There is also an increasing focus on the need for cultural and behavioural changes at the landscape scale to make progress towards delivering sustainability principles (Hobson 2006; Blackstock et al. 2008).

> **Box 1.4** Some key sustainability 'ingredients' applicable to upland areas
>
> - Appropriate scientific knowledge made available to all and involving people at all levels and stages of decision-making, action and evaluation, taking into account all forms of relevant information: scientific, indigenous and local knowledge
> - Conservation and restoration of distinctive and valuable elements of habitats and landscapes
> - Environmental impacts within environmental capacity
> - Environmental justice
> - Improvements in quality of life (reduction of standard of living disparities and poverty): service provision; affordable, well-built housing; high levels of employment and social inclusion
> - Increased innovation, skills and competencies (capacity building)
> - Increased partnership working
> - Independence from financial support through effective use of the market and increased self-reliance
> - Long-term management goals (based on scientific evidence and measuring progress towards goals)
> - Maintained and enhanced quality of primary environmental resources (air, soils, water)
> - Provision of resources and equity for future generations
> - Robust, adaptable and sustainable economies
> - Robust, empowered and sustainable rural communities (developing local solutions)
> - Sustainable travel and tourism
> - Wise use of energy and promotion of waste minimisation and management
> - Working/thinking across boundaries and consider the effects of management – potential/actual – on adjacent/other ecosystems

WHAT CAN WE LEARN FROM COLLABORATIVE, LANDSCAPE-SCALE APPROACHES TO UPLAND GOVERNANCE?

Collaborative and communal approaches have a long history in many of the world's mountain regions (Debarbieux and Price 2008), and an increasingly rich and voluminous body of literature recognises the value of such approaches in helping to deliver sound management and stewardship of uplands at the landscape scale (see Marshall 2005). Collaborative management (or co-management) refers to institutional structures for dialogue and power sharing among resource users and managers. Based on the view that 'sharing authority and decision-making will enhance the process of resource management, making it more responsive to a range of needs' (Castro and Neilson 2001, 231), the concept emphasises social capital and empowerment, echoing the principles of sustainability presented earlier (Shortall 2008). Successful collaborative management is most likely to occur where there is: moderate resource scarcity (Firey 1960); a small and readily identifiable resource with definable boundaries

(Wade 1988); a limited degree of market integration (Agrawal and Yadama 1997); high dependency on the resource for livelihoods (Aralal 2009); and robust local institutions with enforceable rules, adaptive capacity and effective networks (Coleman and Steed 2009). It is rarely possible, however, to identify the individual factors that generate success (Krishna 2003). What is often important is the nature of the process that brings together partners and how they fashion solutions to collaborative management situations within their particular context.

Landscape-scale management is essential to the conservation and management of ecosystem services in working landscapes (MEA 2005). The European Landscape Convention defines a landscape as 'an area as perceived by people, whose character is the result of the action and interaction of natural and/or human factors'. Planning at the 'landscape scale' (Selman 2006) comprises spatial, organisational and temporal dimensions, motivated by a desire to overcome sectoral and single-disciplinary thinking in order to deliver sustainability goals. Good examples include the EU Water Framework Directive, rural-urban fringe plans, and UNESCO Biosphere Reserves. Landscape-scale management requires interaction and co-ordination between landowners and managers and, while often aspired to, tends in practice to be the exception rather than the rule (Selman 2006). In Britain, for example, attempts to encourage collaboration across ownership boundaries at the landscape scale are still somewhat rare (Blackstock et al. 2007) and exacerbated by the predilection for separate sector or agency structures and partnerships to create multiple different spatial boundaries, leading to significant policy 'disintegration' (Scott et al. in press). Disconnections between the scale of management and the scale of ecological processes are also common (Cumming et al. 2006; Mc Morran 2008). In addition, property rights and patterns of landownership further complicate the administration of collaborative ventures (Goldman et al. 2007). Landscape-scale collaboration normally relies on voluntary co-operation without any statutory footing: as a result, commitment to such initiatives can be weak (Prager et al. 2012).

Simultaneously, the concept of spatial planning (Box 1.5) has become the principal planning paradigm across the EU, with a new emphasis on the integration of policies for the development and use of land across scales and across multiple policies and programmes to deliver positive social, economic and environmental outcomes (Albrechts 2004; Tewdwr-Jones et al. 2010). A key element is that of 'place shaping' which may be understood as a collective effort to re-imagine a place in order to define priorities for 'area investment, conservation measures, strategic infrastructure and principles of land use regulation' (Healey 2004, 46).

To mobilise stakeholders within this process requires the development of strategic agendas in diffused power contexts which relies on collaboration and negotiation across many sectors and scales. Such initiatives require the use of partnerships, which are the principal mechanism for the shift in governance style outlined earlier in the chapter (van Huijstee et al. 2007; Derkzen and Bock 2009; Shucksmith 2010). Partnerships are seen as the principal delivery vehicle for the spatial-planning agenda: mechanisms for developing inclusive and deliberative public policy goals (Davidson and Lockwood 2008). Thus, environmental partnerships have emerged in response to

Box 1.5 Upland spatial planning

Effective spatial planning is rare in upland environments because of 'disintegrated' policies that separate planning for urban and rural areas. In Britain, this dualism has led to institutions and structures with different paradigms, mechanisms, agencies, partnerships and tools (Scott et al. in press). Good spatial planning is about connecting across this divide; the following examples demonstrate good practice with relation to uplands:

TAYplan (Strategic Development Plan for Dundee, Angus, Perth and Kinross and Fife, Scotland)

Using the concept of 'nested scales', TAYplan translates intelligence and guidance from global concerns within national Scottish policy to the 'city region' scale. In so doing, the plan connects urban and rural (including upland) land uses through a focus on building partnerships. TAYplan unites four local authorities to integrate land uses through eight policies that signpost visually and spatially what is being done where, providing clear information about where developments/activities can and cannot take place. An 'action plan' outlines practical steps for each policy, clearly stating how it fits with national policy.

Rewilding in the Peak District (northern England)

The Peak District has large areas of moorland with eroding peatland. Heather is managed for grouse but has declined because of overgrazing and poor burning practices. Large areas of bare and exposed peat imply significant losses of carbon to the atmosphere. The National Trust (a conservation organisation) owns land in the area and, to assess the future of rural land use in a way that maximises environmental benefits, conducted a visioning exercise. This successfully improved stakeholder understandings of the value of moorland for societal benefit, enabling support for a strategy that promoted clean water, flood protection and carbon sequestration. Under the banner of 'rewilding', the Trust now takes an experimental approach to allow the development of scrub and to regenerate the peatland. The exercise endorses the key principles of spatial planning in terms of vision, adaptation and market-led tools, taking an opportunity for diversifying from farming of livestock to managing for people (tourism), carbon and clean water as new markets emerge.

sustainability discourse, and the World Summit on Sustainable Development in 2002 formally recognised partnerships as important mechanisms for sustainable development. In an upland context, partnerships tend to include representatives of public, private, voluntary and community interests who are assumed to share a degree of commitment to specific policy objectives. For example, partnerships have served to enable upland stakeholders to recognise their common agendas in rural development, as is the case in the Moors for the Future Partnership in Box 1.6.

Box 1.6 Voluntary collaborative management and partnership working at the landscape scale: The Moors for the Future Partnership (Source: Bonn et al. 2009b)

The Moors for the Future Partnership (MFFP) in the English Peak District (www. moorsforthefuture.org.uk) is a public–private upland partnership between local government, government agencies, non-government organisations, water industries and private landowners. The partnership aims to reverse the degradation of moorland landscape, to improve access for visitors while conserving sensitive areas, and to establish a learning centre to widen people's understanding of how to protect moorlands.

The Peak District uplands provide valuable ecosystem goods and services to local and surrounding populations (Bonn et al. 2009a). The MFFP conducts joint meetings, site visits and specific projects. This has enabled the development of a widely supported land-management strategy with common goals, helped secure funding, enabled collaboration to undertake practical land restoration tasks, and engaged in innovative strategies to deal with conflict situations and environmental risk. Wider awareness raising and engagement beyond the core partners have also been developed to engender a sense of social ownership and responsibility for the landscape.

Partnerships are not problem-free, however; they can often be manipulated by those with power and influence, subsequently distorting outcomes (Edwards et al. 2001; Mackinnon 2002; Vidal 2009). For example, in an assessment of EU LEADER+ partnerships, the key issue was actually how to prevent the programme being 'captured' by the most powerful groups and used to reinforce the status quo (Vidal 2009). Scott et al. (in press) argue that current policy responses trend towards policy disintegration owing to barriers of institutional bias and sectoral thinking that are common in the public, private and voluntary sectors; many stakeholders may seek to preserve their own vested interests in a 'status quo' (Watson 2003; Albrechts 2004). Spatial planning challenges the silo mentality, being set within more joined-up approaches that cross disciplines, professions and sectors. Hence we see emergent policy responses towards catchment- and landscape-based approaches to land-use policy and planning, rather than through the functional management of land within existing administrative boundaries (Selman 2002; International Centre for Protected Landscapes 2008). In partnerships, the various stakeholders differ significantly in terms of their power over, and access to, resources. Thus, property rights play key roles in the sustainable management and improvement of uplands, and have been central to debates on effective governance of natural resources (Grafton 2000).

WHY SHOULD WE AFFORD MORE ATTENTION TO PROPERTY RIGHTS?

In the uplands, the movement away from landscapes of production to landscapes of consumption has significant governance implications for the allocation of property rights. The challenge is to accommodate changing property rights and responsibilities within a range of landownership and management models that foster the long-term provision of livelihoods and ecosystem services from the uplands. The remainder of this chapter considers property rights, both in general and in practice, concluding with an overview of landownership models in Scotland's uplands.

Unpacking private and common property rights

As a relationship, property connects people to one another with respect to land and natural resources (Bromley 1991), and a 'moral perspective' linked to property is increasingly influential in environmental management issues (Brown 2007). The concept of landownership is therefore more complex than it may at first appear (Lachapelle and McCool 2005). There are three key dimensions to unpack. First, ownership has evolved from simple legal and jurisdictional issues of title over land and property to encompass more moral and inclusive notions of community interest and stewardship, expressed within ideas of 'societal good' (Alexander, 1997; Brown 2007). Second, ownership is less about notions of absolute power and control and more about an institutional architecture of regulatory and participatory systems informing decision-making processes and consequential actions. Third, ownership acts across diverse social, political and ecological scales, addressing considerations of 'who' and 'what' are affected by actions, and how plans and decisions are 'owned' spatially. Thus, changes in ownership can expand the number of actors and agencies in property rights issues across different scales and sectors, with increasing intervention in the freedom of an individual to do what she/he wishes (Edwards et al. 2001). Hence property can be viewed as a social relationship that defines the rights of property holders to a particular benefit stream (Mappatoba 2004).

In a private property regime, all rights are held by an individual or organisation (Munton 2009). In modern resource management, private property is distinguished from state property ('state' meaning any level of government), common property held by a defined group usually organised around resource management and cultural continuity, and 'non property' (or 'open access') resources that are not effectively claimed by anyone (but may be used and overused). Property regimes depend on the particular history of a country, or even a region within it, and tend to span four broad types, as shown in Table 1.1.

Five kinds of rights to resources, each of which may be accompanied by obligations or responsibilities and may or may not be transferable, may be identified (Ostrom et al. 2008). Each use of upland resources can be categorised using these distinctions, and considered in terms of the incentives that apply. The five 'positions' which a right holder may occupy are summarised in Box 1.7.

Throughout uplands across the world, there are those who have a private interest in land, for example, for agriculture, forestry, recreation, or mineral extraction. Land is usually purchased privately to pursue production or 'use' rights associated with it (for example, shooting and fishing). This type of ownership can provide and encourage long-term inward investment in resource improvement because it internalises costs

TABLE 1.1 Typology of property rights regimes (Source: Feeny et al. 1990)

Property rights regime	Description
Private property *res private*	Individual has right to appropriate uses of the resource as socially defined; individual has right to exclude others from these uses, and sometimes the right to sell or let the rights to others.
Common property *res communes*	Resource held by a group of users who can exclude non-members and may self-regulate; appropriate uses may be defined by broader society.
State property *res publicae*	Resource rights held by government; government can regulate access and exploitation; public may have access as granted by government.
Open access *res nullius*	Absence of well-defined property rights; resource is unregulated and free to everyone.

Box 1.7 'Positions' of rights holders (Source: Ostrom et al. 2008)

A. **Authorised viewer**, with access but no authority to harvest or make important changes;

B. **Authorised user**, with right to access and withdrawal, usually carefully described in some norm or formalisation;

C. **Claimant**, with rights of access, withdrawal, and management. This status adds security of expectations of long-term capacity to invest and acquire returns on the improvement; the distinction is in a more individualised claim rather than a group claim. It is important that this not be thought of as only something individuals can do or have done.

D. **Proprietor**, who is the holder of rights to access, withdrawal, management, and long-term security of expectations, usually holding substantial obligations to regulate use, investment, and determine access;

E. **Owner**, who is the holder of all the rights, including rights to alienate the resource.

---

and benefits (Demsetz 2002). This is 'often vital to the local economy of mountain areas in Europe', particularly because the 'marginal economics of landowning businesses is often reliant on earnings made elsewhere' (Maxwell and Birnie 2005, 235) or, in many cases in the uplands, on government subsidies. Some express concern about private ownership, however, believing that all too often owners have operated 'extractive regimes, benefiting the owner, failing the environment, doing little for the local community' (Cramb 1996, 13). Similarly, individual interests may not always be compatible with environmental protection (Sandberg 2007). The challenge, therefore, is to design policies and management in ways that overcome this. In Norway, where there is community ownership (with common property rights), with several farmers as commoners and owners, it is argued that the probability of finding a good manager from among the community is better than from a single household (Berge 2002, in Maxwell and Birnie 2005).

The 'commons' have been defined as 'the collective and local ownership of land, resources or ideas, held in an often communal manner, sometimes in opposition to private property' (Holder and Flessus 2008, 300). Common land represents a key component of the upland resource in England and Wales. For example, they account for around 40 per cent of the area of the Lake District, Dartmoor and Brecon Beacons (Scott 1986). The commons (or common land) invoke shared interests and investments in land, culture and nature (Rodgers 1999; Marshall 2005). Their defining characteristics relate to the complex set of institutions and interests that affect their use and management, normally reflecting long-established customary practices and historical traditions (Mitchell 2008). Despite the strong forces for privatisation, common property institutions governing access to, and control over, land and water resources can still be found in many parts of western Europe, located typically in upland or marginal areas (Brown 2006), and also in other mountain areas around the world. Typically, these institutions are for the management and use of grazing land, forests, and irrigation systems, resources for which co-operative regimes bring vital benefits in challenging environments (Debarbieux and Price 2008). In the crofting[3] areas of Scotland's north and west, crofting common grazings are the prevailing form of common land, with some eight hundred administrative units covering 5,000 square kilometres, about 7 per cent of Scotland's area (Edwards 2010). Here, common grazing rights are linked to the tenancy of small individual crofts, unique to the Highlands and Islands of Scotland (Brown 2008).

Land reform and a 'community ownership' turn?

The pursuit of greater economic efficiency and social equity through a redistribution of property rights is a common theme of land reform across various countries, particularly in less developed and transitional economies (Slee et al. 2007). Land reform has been, and continues to be, a significant state-led activity in certain countries, especially post-socialist and post-colonial, in pursuit of wider modernisation and rural development agendas (Bernstein 2002).

Land reform brings into question wider debates on distinctions between statutory and customary law, formal and informal tenure systems, and state-led and community-led reform (Peters 2009). One of the most striking examples is from Zimbabwe, involving the expropriation of land from landowners and its allocation to local people through administrative procedures (Moyo 2011). The role and impacts of these state and developmental influences on customary tenure arrangements for management of the commons remain contentious and poorly understood, however (Upton 2009). In particular, more focus is needed on the nature, processes, and outcomes of initiatives such as land reform and the extent to which they may support, challenge, or undermine existing power structures. This is given detailed consideration in Chapter 3 where the impacts of land reform on private landownership in Scotland's uplands are discussed.

Common property institutions are increasingly proposed as viable communal frameworks for managing resources effectively and productively within 'post-productive' interpretations. The importance of local management in terms of people's rights to 'produce' their own places, rather than have them made for them, has been highlighted (Castree 2003). As Mackenzie (2004, 273) notes, the concept of community plays a key role in re-imagining and reclaiming the land through its place in the collective local imagination. This accords with other claims by indigenous cultural communities as, for example, in Latin America (Griffiths 2004) and customary land tenure in post-apartheid South Africa (Cousins and Claasens 2003; Cousins 2007). In such respects, the role of history and the politics of dispossession provide important themes in cultural cohesion and individual and community reinterpretations of land in Scotland (Rohde 2004; Mackenzie 2006).

Advocates of community-led approaches view community governance of natural resources as desirable because they believe that local groups have higher levels of motivation, knowledge and experience that are inextricably bound up with their livelihoods and community (Evans and Birchenough 2001). Essentially, in such situations, people have to co-operate because they 'depend on each other' (Norberg-Hodge 1991, 46). This reasoning has served as a rallying cry to those dissatisfied with state mismanagement of commons (Smith and Wishnie 2000; Kumar 2005). The example of the Dartmoor Commons is useful here in revealing how, through identification of local problems of overgrazing and pressures for multiple recreation use, a scheme for locally led management was promoted through a private members' bill for regulation of the Dartmoor Commons. The local imperative was favoured as many believed that national legislation would severely damage the livelihoods of the Dartmoor farmers (Scott 1986).

Community ownership and management of land raise important and contested legal and moral notions surrounding definitions of 'community' as a means of bounding collective entitlement (Rohde 2004; Brown 2007; Bryden and Geisler 2007). Some commentators have argued that motivation and collective action increase when land or assets are actually owned by a community, as opposed to being collaboratively managed in conjunction with the state and local actors (Quirk 2007; Slee et al. 2007). Box 1.8 summarises the key benefits of community ownership in comparison to other collaborative or communal arrangements.

There are also several costs and risks, however. In particular, beyond securing the necessary finances required for purchases, the capacity of a community to self-organise to acquire and then manage land-based assets requires additional and ongoing resources and creates liabilities (Dùthchas 2001). Many community acquisitions are undercapitalised and suffer from a lack of appropriate financial and management expertise, which suggests the need for advisory, as well as financial, support (Thake 2006; Quirk 2007). This is reinforced by considerable criticism and scepticism concerning the abilities and capacities of local communities to deliver effectively, as collective action is far more complex and conditional than the theory suggests (Andersson and Gibson 2006; Bouquet 2009; Peters 2009; Upton 2009). These questions are given detailed attention in an examination in Chapter 6 of four community-owned sites in Scotland's uplands.

> **Box 1.8** Benefits of community ownership (Source: Slee et al. 2007, 128)
>
> - Ownership may offer greater security than a tenancy or an informal arrangement for usage, allowing users to plan better for the future.
> - Ownership may offer greater freedom to use an asset for more diverse purposes, opening up opportunities for infrastructure improvements and new production or consumption activities.
> - Ownership may facilitate access to greater funding, through financial gearing/ leverage from a collateral base, thereby enabling more ambitious development.
> - Ownership may encourage social networking that was inaccessible to private or public landlords but which may deliver efficiencies in local service delivery and business development.
> - Ownership may allow more of the surplus (profit) from wealth-creating activities based on the asset to be retained within a community, raising local incomes and employment both directly and indirectly through multiplier effects.
> - Ownership may promote community cohesion and pride through building confidence and a sense of self-worth through control of an asset, particularly if it has iconic, symbolic status that can function as a focal point for community organisation.
> - Ownership, or rather the transition process to ownership and the need to self-organise as an empowered community, may engender a cultural transformation that encourages greater transparency and accountability in decision-making and greater maturity in interacting with other bodies.

Contemporary property rights in the Scottish uplands: a brief overview

The British uplands have multiple uses and users with linked property rights and diverse views on how the uplands are, and should be, managed. Unusually in a European or even global context, the management of these uplands is dominated by sheep farming, game shooting, and forestry undertaken by private landowners and their tenants, with only a few areas retaining common property regimes (Quinn et al. 2010). As briefly mentioned at the beginning of this chapter, land use and management decisions in Scotland are made within the context of a distinct pattern of landownership that has evolved over many centuries: the estates, which are such a characteristic feature of the uplands, can be defined as 'continuous and discrete areas of land held by one owner, whether the owner be an individual, a company, a trust or an institution' (Armstrong and Mather 1983, 9).

Private landownership dominates in the Scottish uplands, and such land is generally regarded as a rural enterprise and/or recreational asset or status symbol for the private landowner, constituting a powerful and complex bundle of property rights (Higgins et al. 2002; Munton 2009). Private ownership often takes the form of a traditional family-owned estate owned as a family or non-family trust, with the head of the family as 'laird' (see Box 1.9). Estates are also owned privately by farmers, commercial developers, and investors seeking capital appreciation, often linked to recreational sporting opportunities. Table 1.2 provides a typology of landowners in the Scottish uplands. Some 'traditional' estates have been owned by the same family for many generations. More than 25 per cent of Scottish landowning families can trace their landowning ancestry back to at least the sixteenth century, and a core of fewer than 1500 privately owned estates has held much of Scotland for nine centuries (Devine 1999). Other estates have been acquired more recently by 'new money', wealthy individuals whose primary residence may be 'down south' in England or overseas. A single private owner may have several land holdings across the country, in a few cases comprising tens of thousands of hectares. Chapters 3, 4 and 5 consider aspects of this type of ownership in more detail.

Box 1.9 Who is the 'laird'?

In Scotland, the traditional term for the owner of an upland estate is the 'laird'. This is an informal title but one redolent of the stereotypical private landowner – privately educated, 'tweed-wearing' and aristocratic, and surrounded by a richly developed set of mythologies, traditions and perceptions (Cramb 1996; Lorimer 2000). Well into the post-war period, the lairds of large estates were generally treated deferentially by local people but times have changed, as later chapters make clear. Old social hierarchies are being broken down, and the 'laird' may now be a conservation organisation or the local community itself. It would be a mistake to equate the title 'laird' to a British 'lord', as it does not confer any political standing, but the fact that some of Scotland's lairds sit in the House of Lords can confuse the outsider.

Twelve per cent of Scotland's land (approximately 930,000 hectares) is owned by public bodies such as Forestry Commission Scotland, local authorities, Scottish Natural Heritage and the Ministry of Defence (Wightman 2010). The Crown Estate owns land in the uplands, notably the 23,000-hectare Glenlivet Estate. Land is also owned by local communities (increasingly since the 2003 land reform legislation, discussed in Chapter 3) and non-governmental organisations (often for biodiversity conservation purposes). These last two ownership models have together been termed 'social ownership' and are considered in more detail in Chapters 6 and 7 respectively.

CONCLUSION

Scotland's uplands represent vital assets at international to local scales and pose important challenges and opportunities for sustainability. One critical component of this is the requirement to improve the governance of these areas by improving the mix of often conflicting plans and their resultant spatial implications for economies,

TABLE 1.2 A typology of landowners in the Scottish uplands (Adapted from: Price et al. 2002). These are not all discrete categories but represent broad types of landowners.

Broad ownership group	Description
Traditional family	Typically hold land that has been in the family for many generations
'Old Money'	Typically bought the land during the expansion of the sporting estates in Victorian times
'New Money'	Wealthy individuals who have bought estates more recently, often investing capital from outwith Scotland
The State	Mainly the lands of Forestry Commission Scotland and other government bodies
Investment owners	Own land primarily as a long-term investment represented by growth in capital values; can be individuals or companies, typically not resident on their land
Farmers	Agricultural owner-occupiers of landholdings of a range of sizes, in the uplands typically grazing sheep or cattle
Commercial developers	Own land for development of industry, mineral exploitation or hydroelectric power, e.g. Rio Tinto Alcan (aluminium smelting using hydroelectric power, Lochaber)
NGO owners	Conservation organisations (e.g. RSPB, National Trust for Scotland, John Muir Trust) who own and manage land with conservation objectives
Community owners	Community bodies that own and manage land for community benefit, e.g. North Harris Trust, Knoydart Foundation

communities and natural resources (Scott et al. in press). Calls for integration are not new and date back to uplands debates in the early 1980s. The presence of power structures resistant to change has limited progress, however, and what has occurred has been characterised by what Curry (1993) calls the fallacy of creeping incrementalism. There are exemplars, however, (for example, the Ben Lawers historical landscape project and the Dumfries and Galloway National Scenic Area management plan, both discussed in Scott 2011) that illuminate a positive direction of travel and highlight the need to mediate and manage conflicts between the values and competing interests of different stakeholders at the earliest possible opportunity.

There will no doubt continue to be a mixed pattern of landownership (and associated distribution of property rights) in Scotland's uplands, and it is important to see this mix as useful and necessary in achieving sustainability in these regions. No single ownership model has all the answers and all have important contributions to make. Recognising the different productive and consumptive claims on upland resources and landscapes, it will also be important for land-use policies to take into account the 'diverse economic conditions, geographical circumstances, sociocultural factors, and traditions to be found there', avoiding 'one size fits all' solutions (Maxwell and Birnie 2005, 237). These issues are explored in the following seven chapters, and the implications for policy and other actions discussed in the final chapter of this book.

NOTES

1. Those goods and services from ecosystems that benefit, sustain and support human liveli-hoods (MEA 2005).
2. See EEA (2010) for an overview of the main European agreements, laws and strategies that drive policy development in Europe's upland areas.
3. A croft is a small land holding, regulated through the Crofting Acts, situated within one of the former crofting counties (Argyll, Inverness-shire, Ross and Cromarty, Sutherland, Caithness, Orkney and Shetland). Crofters constitute around 11 per cent of the population, and 10 per cent of households in Scotland's remote, rural areas.

REFERENCES

Agrawal, A. and Yadama, G. (1997). 'How do local institutions mediate market and population pressures on resources? Forest panchayats in Kumaon, India'. *Development and Change* 28 (3), pp. 435–65.

Albrechts, L. (2004). 'Strategic (spatial) planning re-examined'. *Environment and Planning B: Planning and Design* 31, pp. 743–58.

Alexander, G. S. (1997). 'Civic society'. *Social Legal Studies* 6, pp. 217–34.

Andersson, K. and Gibson, C. C. (2006). 'Decentralized Governance and Environmental Change: Local Institutional Moderation of Deforestation in Bolivia'. *Journal of Policy Analysis and Management* 26 (1), pp. 99–123.

Armstrong, A. S. and Mather, A. S. (1983). *Land Ownership and Land Use in the Scottish Highlands*. Department of Geography, University of Aberdeen.

Berge, E. (2002). *Design Principles of Norwegian Commons*. Proceedings of the Workshop on Future Directions for Common Property Theory and Research, Rutgers University, New Brunswick, 28 February 1997.

Bernstein, H. (2002). 'Land reform: Taking a long(er) view'. *Journal of Agrarian Change* 2, pp. 433–63.

Blackstock, K. L., Brown, K., Davies, B. and Shannon, P. (2007). 'Individualism cooperation and conservation in Scottish farming communities'. In: Cheshire, L., Higgins, V. and Lawrence, G. (eds), *Rural Governance. International Perspectives*. Routledge, London and New York, pp. 191–207.

Blackstock, K. L., White, V., McCrum, G., Scott, A. and Hunter, C. (2008). 'Measuring Responsibility: An Appraisal of a Scottish National Park's Sustainable Tourism Indicators'. *Journal of Sustainable Tourism* 16 (3), pp. 276–97.

Bonn, A., Allott, T., Hubacek, K. and Stewart, J. (eds) (2009). *Drivers of Environmental Change in Uplands*. Routledge, London and New York. 544 pp.

Bonn, A., Allott, T., Hubacek, K. and Stewart, J. (2009a). 'Introduction: Drivers of change in upland environments: concepts, threats and opportunities'. In: Bonn, A., Allott, T., Hubacek, K. and Stewart, J. (eds), *Drivers of Environmental Change in Uplands*. Routledge, London and New York, pp. 1–10.

Bonn, A., Rebane, M. and Reid, C. (2009b). 'Ecosystem services: a new rationale for conservation of upland environments'. In: Bonn, A., Allott, T., Hubacek, K. and Stewart, J. (eds), *Drivers of Environmental Change in Uplands*. Routledge, London and New York, pp. 448–74.

Bouquet, E. (2009). 'State-led Land Reform and Local Institutional Change: Land Titles, Land Markets and Tenure Security in Mexican Communities'. *World Development* 37 (8), pp. 1390–9.

Brooker, R. (2011). 'The changing nature of Scotland's uplands – an interplay of processes and timescales', In: S. J. Marrs, S. Foster, C. Hendrie, E. C. Mackey and D. B. A. Thompson (eds), *The Changing Nature of Scotland*. The Stationery Office, Edinburgh, pp. 381–96.

Brown, K. M. (2007). 'Reconciling moral and legal collective entitlement: Implications for community-based land reform'. *Land Use Policy* 24 (4), pp. 633–43.

Brown, A. P. (2008). 'Crofter, forestry land reform and the ideology of community'. *Social and Legal Studies* 17 (3), pp. 333–49.

Bryden, J. and Geisler, C. (2007). 'Community-based land reform: Lessons from Scotland'. *Land Use Policy* 24 (1), pp. 24–34.

Bunce, R. G. H., Barr, C. J., Clarke, R. T. and Howard, D. C. (1996). 'Land classification for strategic ecological survey'. *Journal of Environmental Management* 79, pp. 63–77.

Burt, T. P. (2001). 'Integrated management of sensitive catchment systems'. *Catena* 42 (2–4), pp. 275–90.

Carling, P. A., Irvine, B. J., Hill, A. and Wood, M. (2001). 'The efficacy of soil conservation practices in UK upland forestry: a review'. *Science of the Total Environment* 265, pp. 209–27.

Carter, C. (2006). *Environmental Governance: The Power and Pitfalls of Participatory Processes.* The Macaulay Institute, Aberdeen.

Castree, N. (2003). 'Differential geographies: place, indigenous rights and "local" resources'. *Political Geography* 23, pp. 133–67.

Castro, A. P. and Nielson, E. (2001). 'Indigenous people and co-management: Implications for conflict management'. *Environmental Science and Policy* 4, pp. 229–39.

Coleman, E. A and Steed, B. C. (2009). 'Monitoring and sanctioning in the commons: An application to forestry'. *Ecological Economics* 88, pp. 2106–13.

Cousins, B. (2007). 'More than socially embedded: The distinctive character of "communal tenure" regimes in South Africa and its implications for land policy'. *Journal of Agrarian Change* 7 (3), pp. 281–315.

Cousins, B. and Claassens, A. (2003). 'Communal tenure systems in South Africa: Past, present and future'. *Development Update* 4, pp. 55–78.

Cramb, A. (1996). *Who owns Scotland now? The use and abuse of private land.* Mainstream, Edinburgh. 208 pp.

Cumming, G. S., Cumming, D. H. M., Redman, C. L. (2006). 'Scale mismatches in social-ecological systems: causes, consequences, and solutions'. *Ecology and Society* 11, 14.

Curry, N. R. (1993). *Countryside planning: A look back in anguish.* Unpublished Inaugural Lecture, Cheltenham and Gloucester College of Higher Education, Cheltenham.

Curry, N. R. (2009). 'Leisure in the landscape: rural incomes and public benefits'. In: Bonn, A., Allott, T., Hubacek, K. and Stewart, J. (eds), *Drivers of Environmental Change in Uplands.* Routledge, London and New York, pp. 277–90.

Davidson, J. and Lockwood, M. (2008). 'Partnerships as Instruments of Good Regional Governance: Innovation for Sustainability in Tasmania?' *Regional Studies* 42 (5), pp. 641–56.

Debarbieux, B. and Price, M. F. (2008). 'Representing mountains: From local and national to global common good'. *Geopolitics* 13, pp. 148–68.

Demsetz, H. (2002). 'Toward a theory of property rights II: the competition between private and collective ownership'. *The Journal of Legal Studies* 31, S653–S672.

Derkzen, P. and Bock, B. (2009). 'Partnership and role perception, three case studies on the meaning of being a representative in rural partnerships'. *Environment and Planning C* 27, pp. 75–89.

Devine, T. M. (1999). *The Scottish Nation 1700–2000.* Allen Lane, London. 696 pp.

Dùthchas (2001). 'Area Sustainability Strategies for Peripheral Rural Areas: January 1998–April 2001. Dùthchas – The Final Report'. <http://www.duthchas.org.uk/frameset1.html> (last accessed 2 July 2009).

Edwards, T. (2010). 'Crofting Reform (Scotland) Bill'. SPICe Briefing 10/01. Scottish Parliament Information Centre, Edinburgh.

Edwards, W. J., Goodwin, M., Pemberton, S. and Woods, M. (2001). 'Partnerships, power, and scale in rural governance'. *Environment and Planning C* 19, pp. 289–310.

Eiter, S. and K. Potthoff (2007). 'Improving the factual knowledge of landscapes: Following up the European Landscape Convention with a comparative historical analysis of forces of landscape change in the Sjodalen and Stolsheimen mountain areas, Norway'. *Norsk Geografisk Tidsskrift – Norwegian Journal of Geography* 61, pp. 145–56.

European Environment Agency (2010). *Europe's ecological backbone: recognising the true value of our mountains.* EEA Report No 6/2010.

Evans, S. M. and Birchenough, A. C. (2001). 'Community-based management of the environment: lessons from the past and options for the future'. *Aquatic Conservation* 11, pp. 137–47.

Fazey, I. (2010). 'Resilience and Higher Order Thinking'. *Ecology and Society* 15, 22.

Feeny, D., Berkes, F., McCay, B. J. and Acheson, J. M. (1990). 'The tragedy of the commons: twenty-two years later'. *Human Ecology* 18 (1), pp. 1–19.

Firey, W. (1960). *Man, Mind and Land.* Glencoe Free Press.

Goldman, R. L., Thompson, B. H. and Daily, G. C. (2007). 'Institutional incentives for managing the landscape: inducing cooperation for the production of ecosystem services'. *Ecological Economics* 64, pp. 333–43.

Grafton, R. Q. (2000). 'Governance of the commons: a role for the state?' *Land Economics* 76, pp. 504–17.

Grêt-Regamey, A., Brunner, S. H. and Kienast, F. (2012). 'Mountain Ecosystem Services: Who Cares?' *Mountain Research and Development* 32 (S1), S23–S34.

Griffiths, T. (2004). 'Indigenous peoples, land tenure and land policy in Latin America'. < http://www.fao.org/docrep/007/y5407t/y5407t0a.htm> (last accessed 8 January 2013).

Hart, K., Baldock, D., Weingarten, P., Osterburg, B., Povellato, A., Vanni, F., Pirzio-Biroli, C. and Boyes, A. (2011). 'What tools for the European Agricultural Policy to encourage the provision of public goods?' Directorate General for Internal Policies – Policy Department B: Structural and Cohesion Policies. European Parliament, Brussels.

Healey, P. (2004). 'The treatment of space and place in the new strategic spatial planning in Europe'. *International Journal of Urban and Regional Research* 28 (1), pp. 45–67.

Hellman, F. and Verburg, P. H. (2010). 'Impact assessment of the European biofuel directive on land use and biodiversity'. *Journal of Environmental Management* 91 (6), pp. 1389–96.

Higgins, P., Wightman, A. and Macmillan, D. (2002). *Sporting Estates and Recreational Land Use in the Highlands and Islands of Scotland.* ESRC, Swindon.

Hobson, K. (2006). 'Environmental responsibility and the possibilities of pragmatist-orientated research'. *Social and Cultural Geography* 7 (2), pp. 283–98.

Holder, J. B and Flessus, T. (2008). 'Emerging Commons'. *Social and Legal Studies* 17 (3), pp. 299–310.

van Huijstee, M. M., Francken, M. and Leroy, P. (2007). 'Partnerships for sustainable development: a review of current literature'. *Environmental Sciences* 4 (2), pp. 75–89.

International Centre for Protected Landscapes (2008). *Identifying Good Practice from countries implementing the European Landscape Convention.* Project Ref: ICP/001/07, International Centre for Protected Landscapes, Aberystwyth.

Ives, J. D., Messerli. B. and Spiess, E. (1997). 'Mountains of the world – A global priority'. In: B. Messerli and J. D. Ives (eds), *Mountains of the World: A Global Priority.* Parthenon, New York and London, pp. 1–15.

Kates, R. W., Parris, T. M. and Leiserowitz, A. A. (2005). 'What is sustainable development? Goals, Indicators, Values, and Practice'. *Environment* 47(3), pp. 8–21.

Körner, C., Ohsawa, M., Spehn, E., Berge, E., Bugmann, H., Groombridge, B., Hamilton, L., Hofer, T., Ives, J., Jodha, N., Messerli, B., Pratt, J., Price, M., Reasoner, M., Rodgers, A., Thonell, J. and Yoshino, M. (2005). 'Mountain systems'. In: Hassan, R., Scholes, R. and Ash, N. (eds), *Ecosystems and Human Well-being: Current State and Trends*, Volume 1. Millennium Ecosystem Assessment, Island Press, Washington, DC, pp. 681–716.

Krishna, A. (2003). 'Partnerships between local governments and community based organisations: exploring the scope for synergy'. *Public Administration and Development* 23, pp. 361–71.

Kumar, C. (2005). 'Revisiting "community" in community-based natural resource management'. Community Development Journal 40 (3), pp. 275–85.

Lachapelle, P. R. and McCool, S. F. (2005). 'Exploring the Concept of "Ownership" in Natural Resource Planning'. *Society and Natural Resources* 18 (3), pp. 279–85.

Lorimer, H. (2000). 'Guns, game and the grandee: the cultural politics of deerstalking in the Scottish Highlands'. *Ecumene* 7 (4), pp. 403–31.

Luke, T. W. (2005). 'Neither sustainable nor development: reconsidering sustainability in development'. *Sustainable Development* 13, pp. 228–38.

Mackenzie, A. F. D. (2004). 'Re-imagining the land, North Sutherland, Scotland'. *Journal of Rural Studies* 20, pp. 273–87.

Mackenzie, A. F. D (2006). 'A working land: crofting communities, place and the politics of the possible in post-Land Reform Scotland'. *Transactions of the Institute of British Geographers* 31 (3), pp. 383–98.

MacKinnon, D. (2002). 'Rural governance and local involvement: assessing state–community relations in the Scottish Highlands'. *Journal of Rural Studies* 18, pp. 307–24.

Mc Morran, R. (2008). 'Scale mis-matches in social–ecological systems: a case study of multifunctional forestry in the Cairngorms region of Scotland'. *Aspects of Applied Biology* 85, pp. 41–8.

McVittie, A., Moran, D., Smyth, K. and Hall, C. (2005). *Measuring public preferences for the uplands. Final Report to the Centre for the Uplands, Cumbria*. Scottish Agricultural College, Edinburgh.

Mansfield, L. (2011). *Upland Agriculture and the Environment*. Badger Press, Bowness-on-Windermere. 360 pp.

Mappatoba, M. (2004). *Co-Management of Protected areas: the Case of Community Agreements on Conservation in the Lore Lindu National Park Central Sulawesi Indonesia*. Institute of Rural Development: University of Gottingen, Germany.

Marrs, S. J., Foster, S., Hendrie, C., Mackey, E. C. and Thompson, D. B. A. (eds) (2011). *The Changing Nature of Scotland*. The Stationery Office, Edinburgh. 528 pp.

Marshall, G. R. (2005). *Economics for Collaborative Environmental Management*. Earthscan, London. 184 pp.

Maselli, D. (2012). 'Promoting Sustainable Mountain Development at the Global Level. *Mountain Research and Development* 32 (S1), S64–S70.

Mather, A. S., Hill, G. and Nijnik, M. (2006). 'Post-productivism and rural land use: cul de sac or challenge for theorization?' *Journal of Rural Studies* 22, pp. 441–55.

Maxwell, J. and Birnie, R. (2005). 'Multi-purpose management in the mountains of Northern Europe – policies and perspectives'. In: Thompson, D. B. A., Price, M. F. and Galbraith, C. A. (eds), *Mountains of Northern Europe: Conservation, Management, People and Nature*. TSO Scotland, Edinburgh, pp. 227–38.

Maxwell, T. J. and Cannell, M. G. R. (2000). 'The environment and land use of the future'. In: Holmes, G. and Crofts, R. (eds), *Scotland's Environment: the future*. Tuckwell Press, East Linton, pp. 30–51.

MEA (Millennium Ecosystem Assessment), 2005. Ecosystems and Human Well Being. Island Press, Washington, DC.

Meadowcroft, J. (2000). 'Sustainable development: a new(ish) idea for a new century?' *Political Studies* 48, pp. 370–87.

Midgley, A., Williams, F., Slee, B. and Renwick, A. (2008). *Primary Land-Based Business Study*. Scottish Agricultural College, Edinburgh.

Mitchell, J. (2008). 'What public presence? Access, common and property rights'. *Social and Legal Studies* 17 (3), pp. 351–67.

Mowle, A. (1997). 'The managing of the land'. In: Magnusson, M. and White, G. (eds), *The Nature of Scotland: Landscape, Wildlife and People*. Canongate, Edinburgh, pp. 133–43.

Moyo, S. (2011). 'Three decades of agrarian reform in Zimbabwe'. *Journal of Peasant Studies* 38, pp. 493–531.

Munton, R. (2009). 'Rural Landownership in the United Kingdom'. *Land Use Policy* 26S, S54–S61.

Newell, P. and Frynas, J. G. (2007). 'Beyond CSR? Business, poverty and social justice: an introduction'. *Third World Quaterly* 28 (4), pp. 669–81.

Norberg-Hodge, H. (1991). *Ancient Futures, Learning from Ladakh*. Rider, London. 240 pp.

Norton, B. G. (2005). *Sustainability. A Philosophy of Adaptive Ecosystem Management*. University of Chicago Press, Chicago. 608 pp.

Olsson, G. A. (2005). 'The use and management of Norwegian mountains reflected in biodiversity values – what are the options for future food production?' In: Thompson, D. B. A., Price, M. F. and Galbraith, C. A. (eds), *Mountains of Northern Europe: Conservation, Management, People and Nature*. The Stationery Office, Edinburgh, pp. 151–62.

Ostrom, E. (2009). 'Design Principles of Robust Property Rights Institutions: What have We Learned?' In: Ingram, G. K. and Hong, Y.–H. (eds), *Property Rights and Land Policies*. Lincoln Institute of Land Policy, Cambridge, MA, pp. 25–51.

Ostrom, E., Burger, J., Field, C. B., Norgaard, R. B. and Policansky, D. (1999). 'Revisiting the commons: local lessons, global challenges'. *Science* 284, PP. 278–82.

Perry, B. (2007). 'The Multi-level Governance of Science Policy in England'. *Regional Studies* 41 (8), pp. 1051–67.

Peters, P. E. (2009). 'Challenges in Land Tenure and Land Reform in Africa: Anthropological Contributions'. *World Development* 37, pp. 1317–25.

Prager, K., Reed, M. S. and Scott, A. J. (2012). 'Viewpoint: Encouraging collaboration for the provision of ecosystem services at a landscape scale – Rethinking agri-environmental payments'. *Land Use Policy* 29, pp. 244–9.

Price, M. F. and Kim, E. G. (1999). 'Priorities for sustainable mountain development in Europe'. *International Journal of Sustainable Development and World Ecology* 6, pp. 203–19.

Price, M. F., Dixon, B. J., Warren, C. R. and Macpherson, A. R. (2002). *Scotland's Mountains: Key Issues for their Future Management*. Scottish Natural Heritage, Battleby. 90 pp.

Price, M. F. and Kohler, T. (2013). 'Sustainable mountain development'. In: Price, M. F., Byers, A. C., Friend, D. A.. Kohler, T. and Price, L. W. (eds), *Mountain Geography: Physical and Human Dimensions*. University of California Press, Berkeley, in press.

Quinn, C. H., Huby, M., Kiwasila, H. and Lovett, J. C. (2007). 'Design principles and common pool resource management: an institutional approach to evaluating community management in semi-arid Tanzania'. *Journal of Environmental Management* 84, pp. 100–13.

Quinn, C. H., Fraser, E. D. G., Hubacek, K. and Reed, M. S. (2010). 'Property rights in UK uplands and the implications for policy and management'. *Ecological Economics* 69, pp. 1355–63.

Quirk, B. (2007). 'Making assets work'. The Quirk Review of community management and ownership of public assets. HMSO, London.

Ratcliffe, D. A. (ed.) (1977). *A Nature Conservation Review*. Two volumes. Cambridge University Press, Cambridge.

Ratcliffe, D. A. and Thompson, D. B. A. (1988). 'The British uplands: their ecological character and international significance'. In: Usher, M. B. and Thompson, D. B. A. (eds), *Ecological Change in the Uplands*. Blackwell, Oxford, pp. 9–36.

Ray, C. (2006). 'Neo-endogenous rural development in the EU'. In: Cloke, P., Marsden, T. and Mooney, P. (eds), *Handbook of Rural Studies*. Sage, London, pp. 278–91.

Reed M. S., Bonn A., Slee W., Beharry-Borg N., Birch J., Brown I., Burt T. P., Chapman D., Chapman P. J., Clay G., Cornell S. J., Fraser E. D. G., Glass J. H., Holden J., Hodgson J. A., Hubacek K., Irvine B., Jin N., Kirkby M. J., Kunin W. E., Moore O., Moseley D., Prell C., Price

M. F., Quinn C., Redpath S., Reid C., Stagl S., Stringer L. C., Termansen M., Thorp S., Towers W. and Worrall F. (2009). 'The future of the uplands'. *Land Use Policy* 26S, S204–S216.

Reed, M. S., Buckmaster, S., Moxey, A. P., Keenleyside, C., Fazey, I., Scott, I., Thomson, K., Thorp, S., Anderson, R., Bateman, I., Bryce, R., Christie, M., Glass, J., Hubacek, K., Quinn, C., Maffey, G., Midgely, A., Robinson, G., Stringer, L. C., Lowe, P. and Slee, R. (2011). 'Policy Options for Sustainable Management of UK Peatlands'. IUCN Technical Review 12. IUCN UK Peatland Programme, Edinburgh.

Rodgers, C. P. (1999). 'Environmental management of common land: towards a new legal framework.' *Journal of Environmental Law* 11 (2), pp. 231–55.

Rohde, R. (2004). 'Ideology, Bureaucracy and Aesthetics: Landscape Change and Land Reform in Northwest Scotland'. *Environmental Values* 13, pp. 199–221.

Rye, J. F. and Berg, N. G. (2011). 'The second home phenomenon and Norwegian rurality'. *Norsk Geografisk Tidsskrift – Norwegian Journal of Geography* 65, pp. 126–36.

SAC (2008). *Farming's retreat from the hills.* Scottish Agricultural College Rural Policy Centre, Edinburgh.

Sandberg, A. (2007). 'Property rights and ecosystem properties'. *Land Use Policy* 24, pp. 613–23.

Satsangi, M. (2009). 'Community Land Ownership, Housing and Sustainable Rural Communities'. *Planning Practice and Research* 24 (2), pp. 251–62.

Schild, A. and Scharma, E. (2011). 'Sustainable Mountain Development Revisited'. *Mountain Research and Development* 31 (3), pp. 237–41.

Scott, A. J. (1986). *Issues in Common Land Management: A Case Study of the Dartmoor Commons.* Unpublished PhD thesis. University of Wales, Aberystwyth.

Scott, A. J. (2011). 'Beyond the conventional: Meeting the challenges of landscape governance within the European Landscape Convention?' *Journal of Environmental Management* 92, pp. 2754–62.

Scott, A. J. (2012). 'Exposing, Exploring and Navigating the built and natural divide in public policy and planning'. *In Practice* magazine (March 2012). Institute of Ecology and Environmental Management, pp. 20–3.

Scott, A. J., Shorten, J., Owen, R. and Owen, I. G. (2009). 'What kind of countryside do we want: perspectives from Wales, UK'. *Geojournal* doi 10.1007/s10708-009-9256-y.

Scott, A. J., Carter, C. E., Larkham, P., Reed, M., Morton, N., Waters, R., Adams, D., Collier, D., Crean, C., Curzon, R., Forster, R., Gibbs, P., Grayson, N., Hardman, M., Hearle, A., Jarvis, D., Kennet, M., Leach, K., Middleton, M., Schiessel, N., Stonyer, B. and Coles, R. (in press) 'Disintegrated Development at the Rural Urban Fringe: Re-connecting spatial planning theory and practice'. *Progress in Planning.*

Selman, P. (2002). 'Multi-function Landscape Plans: a missing link in sustainability planning?' *Local Environment* 7 (3), pp. 283–94.

Selman, P. (2006). *Planning at the Landscape Scale.* Routledge, London. 213 pp.

Shortall, S. (2008). 'Are rural development programmes socially inclusive? Social inclusion, civic engagement, participation, and social capital: Exploring the differences'. *Journal of Rural Studies* 24 (4), pp. 450–7.

Shucksmith, M. (2000). 'Endogenous development, capacity building and inclusion: perspectives from the UK experience of LEADER'. *Sociologia Ruralis* 40 (2), pp. 208–18.

Shucksmith, M. (2010). 'Disintegrated Rural Development? Neo-endogenous Rural Development, Planning and Place-Shaping in Diffused Contexts'. *Sociologia Ruralis* 50 (1), pp. 1–14.

Skuras, D., Petrou, A. and Clark, G. (2007). 'Demand for rural tourism: the effects of quality and information'. *Agricultural Economics* 35 (2), pp. 183–92.

Slee, W. (2005). 'From countrysides of production to countrysides of consumptions?' *Journal of Agricultural Science* 143, pp. 255–65.

Slee, B., Blackstock, K., Brown, K. M., Moxey, A., Cook, P. and Greive, J. (2007). 'Monitoring and evaluating the impacts of land reform'. Report for the Scottish Government, September 2007.

Smith, E. A. and Wishnie, M. (2000). 'Conservation and subsistence in small scale societies'. *Annual Review of Anthropology* 29, pp. 493–524.

Tewdwr-Jones, M., Gallent, N. and Morphet, J. (2010). 'An Anatomy of Spatial Planning: Coming to Terms with the Spatial Element in UK Planning'. *European Planning Studies* 18 (2), pp. 239–57.

Thake, S. (2006). *Community Assets: the Benefits and Costs of Community Management and Ownership.* Department of Communities and Local Government, HMSO, London.

Thompson, D. B. A., Nagy, L., Johnson, S. M. and Robertson, P. (2005). 'The nature of mountains: an introduction'. In: Thompson, D. B. A., Price, M. F. and Galbraith, C. A. (eds), *Mountains of Northern Europe: Conservation, Management, People and Nature.* The Stationery Office, Edinburgh, pp. 43–55.

Tovey, H. and Mooney, R. (2006). *Sixth Framework Programme Priority 7: Citizens and Governance in a Knowledge-Based Society.* Final Report. CORASON Project, EU.

Upton, C. (2009). '"Custom" and Contestation: Land Reform in Post-Socialist Mongolia'. *World Development* 37 (8), pp. 1400–10.

Vidal, R. V. V. (2009). 'Rural development within the EU LEADER+ programme: new tools and technologies'. *AI and Society* 23, pp. 575–602.

Wade, R. (1988). *Village Republics: Economic Conditions for Collective Action in South India.* Cambridge University Press, Cambridge. 256 pp.

Warren, C. R. (2009). *Managing Scotland's Environment.* Second Edition. Edinburgh University Press, Edinburgh. 432 pp.

Watson, V. (2003). 'Conflicting rationalities: implications for planning theory and ethics'. *Planning Theory and Practice* 4 (4), pp. 395–407.

Werritty, A. (2002). 'Living with uncertainty: climate change, river flows and water resource management in Scotland'. *Science of the Total Environment* 294 (1–3), pp. 3–11.

Wightman, A. (2010) *The Poor Had No Lawyers: Who owns Scotland and how they got it.* Birlinn, Edinburgh. 320 pp.

World Commission on Environment and Development (1987). *Our Common Future.* Oxford Paperbacks, Oxford. 416 pp.

CHAPTER TWO

Recognising Scotland's upland ecosystem services

Jayne Glass, Martin F. Price, Alister Scott, Charles Warren and Robert Mc Morran

INTRODUCTION

Uplands provide and support a diverse range of ecosystem services that underpin the social fabric and economies of associated rural communities, and also contribute to the well-being and economies of urban areas and populations (Reed et al. 2009). The uplands cover some 18 per cent of Britain; mountains, upland moors and heaths represent the country's largest unfragmented semi-natural habitats (van der Wal et al. 2011). The majority of these occur in Scotland (3.4 million hectares) where they comprise 43 per cent of the land area (van der Wal et al. 2011). Thus, the Scottish uplands are increasingly recognised as a core source of crucial ecosystem services (Figure 2.1). This chapter focuses on Scotland's uplands and uses ecosystem services as the lens within which to highlight the value these uplands provide to society.

Today's upland landscapes have been fashioned by the interactions of people and nature over a long period of time (Brooker 2011); virtually all of the uplands are actively managed as multifunctional, cultural landscapes, kept in an 'open state' by practices such as grazing, cutting and burning (Ratcliffe and Thompson 1988; Dodgshon and Olsson 2006; Lindsay and Thorp 2011). Scotland's uplands are both valuable and vulnerable, and have important roles to play in delivering sustainability principles, as policy documents have repeatedly recognised. For example, in 1990, the Countryside Commission for Scotland (now Scottish Natural Heritage, SNH) recognised the 'need to give new impetus to the care of Scotland's mountains and also to review the role that these areas play for a rapidly changing society' (Countryside Commission for Scotland 1990). This was reiterated in 1996 in the 'Scottish agenda for sustainable mountain development' which included eight recommendations covering environmental, economic and social aspects of the uplands (Wightman 1996: see Box 2.1). In 2008, the Royal Society of Edinburgh's (RSE) Hills and Islands Inquiry emphasised the importance of 'delivering economic viability and employment opportunities, social benefits and the care and enhancement of the natural resource' (McCrone et al. 2008, 26). In 2009, SNH reviewed and updated the series of 'Natural Heritage Futures'

FIGURE 2.1 *The Scottish uplands: selected ecosystem services*

documents, first published in 2002, which included a vision of how the natural heritage of Scotland's hills and moors could look in 2025, based on sustainable use of natural resources (Jordan and Christie 2010). Strikingly, all of these documents recognised the need for: integration of management objectives; greater public participation in setting goals; recognition of the value of outdoor recreation; enhanced protection of the natural heritage, and mechanisms to ensure environmental restoration. With regard to the uplands, particular changes in management pressures recognised since 2002 were 'an increased focus on the links between land management and climate change', afforestation, renewable energy developments, and changes in livestock grazing patterns, with decreases in the numbers of hill sheep and cattle (Jordan and Christie 2010: 80).

The recent United Kingdom National Ecosystem Assessment (UK NEA 2011) analysed the relative importance of eight 'Broad Habitats' in delivering ecosystem services. Of these habitats, 'Mountains, Moorlands and Heaths' (MMH) cover the uplands; Figure 2.2 shows both the importance of MMH for delivering a range of services and the 'direction of change in the flow of each service'. It is interesting to note that none of the services has been labelled as 'improving', with deterioration observed in peatlands, livestock/aquaculture (livestock numbers are discussed below), hazard (*sic*), and pollination.

Box 2.1 Eight measures for a 'Scottish agenda for sustainable mountain development' (Source: Wightman 1996)

1. Stronger controls over land use and better land-use integration in mountain country as a whole;
2. Maximum protection for the most valuable areas;
3. Expansion of public ownership of the most valuable mountain country;
4. Enhanced participation by, and benefits to, mountain communities;
5. Strategic planning for outdoor recreation;
6. Commitment to restore mountain ecosystems;
7. Improved research and monitoring of mountain areas;
8. International collaboration in montane conservation.

This chapter discusses the key drivers (driving forces) affecting the management of upland ecosystem services in Scotland: changes in 'traditional' industries (particularly agriculture, forestry and field sports); growth in consumptive land uses (including recreation and tourism, as discussed in Chapter 1); energy and climate change (including renewable energy); environment regulation and protection; and the dynamics of upland economies and communities. Particular attention is paid to those drivers that are affected by the actions of landowners and managers within Scotland's upland estates, as introduced in Chapter 1.[1]

FROM SECTORAL TO INTEGRATED APPROACHES TO MANAGING UK UPLANDS

In the United Kingdom, upland land-management policies have traditionally been sectoral (Mowle 1997; Thompson et al. 2005), set within what Scott et al. (2004) term 'institutional myopia'. This was demonstrated particularly during the British foot-and-mouth disease outbreak in 2001, with a disintegrated approach to tackling the outbreak evident in England and Wales (Scott et al. 2004). There are, however, important signals that reflect a shift from sectoral thinking towards more integrated and multifunctional approaches. For example, contemporary agricultural policy has seen a shift towards post-productivism, reflecting the intersection of production and consumption functions in the same geographical space (as discussed in Chapter 1); cultural values of community and tradition are reinforced by touristic visions of a cultural idyll (Shaw and Whyte 2008); and ecological concerns over biodiversity are now tackled through landscape-scale approaches, using partnerships as the primary delivery vehicle (Prager et al. 2012; see Chapter 1). Box 2.2 considers Scotland's Moorland Forum as an example of collaboration around the delivery of joint objectives for moorland environments.

In Scotland, rural land-use policy between 1950 and 1980 was persistently criticised for being sectoral and for showing a lack of integration and joined-up thinking at all levels of governance (Lowe et al. 1986; Davidson 1994; Crofts 2000). The primacy of agriculture and forestry trumped wider rural-development objectives, limiting

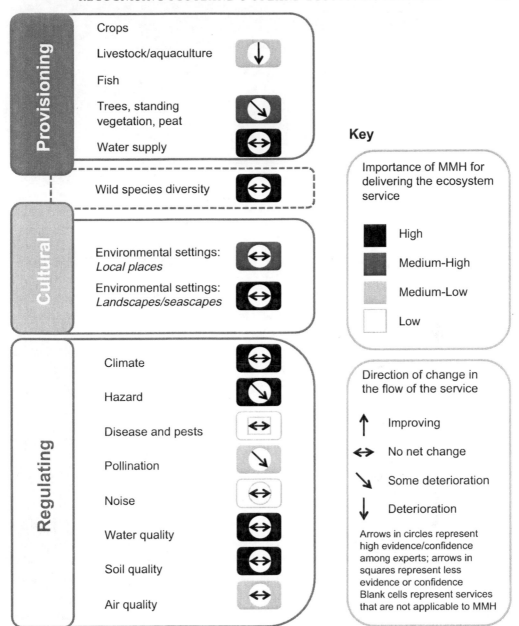

FIGURE 2.2 *Relative importance of Mountains, Moorlands and Heaths (MMH) in delivering ecosystem services, and overall direction of change in service flow since 1990 (Adapted from: UK NEA 2011: 11)*

the effective delivery of rural policy (Scott et al. 2007). Consequently, in Scotland from the 1990s, a more integrated approach to policy-making has increasingly been taken (Scottish Executive 2006; Warren 2009), with moves being made towards multifunctionality (Midgley et al. 2008) and landscape-scale projects (particularly through the United Kingdom's signature of the European Landscape Convention

---

Box 2.2 Scotland's Moorland Forum – landscape-scale governance?
(Adapted from: Lindsay and Thorp 2011)

Scotland's Moorland Forum was formed in 2002 as a partnership that aims to 'sustain and enhance the extent, delivery and range of habitats, species and enterprises encompassing moorland' (Lindsay and Thorp 2011, 417). There is a particular focus on reversing the loss of heather cover (a quarter has been lost since the 1940s, and losses continue at an estimated rate of 0.5 per cent each year).

The Forum consists of thirty member organisations that represent a range of interests, including public authorities, academic institutions and representative bodies. The forum seeks input from its members on key issues related to Scotland's uplands, and works to find solutions that are mutually acceptable within the group. A number of working groups allow the development and delivery of a work programme that considers forestry, agriculture, carbon, access and awareness, and muirburn (the burning of heather).

---

in 2006: International Centre for Protected Landscapes 2008). The concept of sustainability is also now embedded within the Scottish Government's National Performance Framework (Scottish Government 2012a) and 'Ecosystem Services' is one of the Scottish Government's major research themes within the Rural Affairs and Environment portfolio (Scottish Government 2011a).

An integrated approach is also found to some extent within the current Scotland Rural Development Programme (SRDP 2007–13), which provides £1.6 billion of outcome-oriented funding to farmers, communities, land managers and landowners for measures to achieve social and economic goals as well as environmental ones (Scottish Government 2012b). As the delivery mechanism for the EU's Common Agricultural Policy (CAP) in Scotland, the influence of the SRDP on land use and management is significant, comprising a unified support scheme encompassing farming, forestry, primary processing, diversification, rural enterprise, tourism and business development. The integration of all these previously separate funding streams is a notable 'first', as is the incorporation of explicit landscape-scale objectives and outcomes (Scott 2011).

Integrated, joined-up management is now almost universally perceived as both necessary and desirable, not only to ensure the delivery of effective, sustainable environmental management and coherent policy-making but also to avoid wasteful duplication and conflict. Consequently, it is repeatedly championed in public policy pronouncements (exemplified above), and is a central rationale underlying the Scottish Land Use Strategy (Box 2.3), designed as an overarching 'umbrella of strategies'. Integrated land use is neither easy nor quick to achieve, however. Indeed, there are those who doubt whether the Scottish Government's ambition of a fully 'joined-up' rural policy is a feasible objective (Jordan and Halpin 2006). Nor is integrated land use easy to find in practice, not least because of the legacy of decades of resolutely sectoral policy-making. Government departments, both centrally and locally, have long had a strong preference for single-issue policies and limited timeframes (Allmendinger

2003). Only on some large private estates has there been a long-standing tradition of practising integrated land use with multiple objectives over multigenerational time-scales. Thus, notwithstanding the many criticisms of private ownership (see Chapter 3), at its best, estate management provides an exemplar of integrated, cross-sectoral management of the upland environment (Selman 1988; Kerr 2004).

Box 2.3 Scotland's Land Use Strategy (Sources: Scottish Government 2011b; Scottish Government 2011c)

The Scottish Land Use Strategy (LUS), laid before the Scottish Parliament in 2011, resulted from a commitment of the Climate Change (Scotland) Act (2009). The LUS aims to support the sustainable and optimal use of Scotland's land resource with a long-term strategic vision of: 'a Scotland where we fully recognise, understand and value the importance of our land resources, and where our plans and decisions about land use deliver improved and enduring benefits, enhancing the wellbeing of our nation'. Three key 'shifts in approach' are envisioned as central to the implementation of the LUS through public bodies (and public sector policies) and more widely: delivering multiple benefits; partnerships with nature; and linking people with the land. The vision is linked with three core objectives, which can be summarised as:
- land-based businesses working with nature;
- responsible stewardship of Scotland's natural resources;
- urban and rural communities better connected to the land and positively influencing land use.

These objectives are linked with ten key principles, which can be summarised as:
- land use for multiple benefits;
- light-touch regulation and efficient cost-effective incentives;
- the use of the most suitable land for the most suitable purposes;
- greater understanding of ecosystem services;
- sympathetic management of landscape change;
- greater understanding of climate change;
- restoration of derelict land;
- encouragement of recreational opportunities;
- provision of opportunities for the public to contribute to debates and decisions about land;
- increased understanding of links between land use and daily living.

A number of proposals for action have also been developed, including the development of an action plan, annual progress reviews and five-yearly revisions, and the identification of suitable demonstration projects for strategy implementation. These proposals also include the alignment of the regulatory and incentives frameworks for land use with the LUS. Critically, the LUS is not a prescriptive strategy for all land parcels but rather a long-term strategic process which provides the context for all land-use decisions and relevant policies in Scotland and a forum for dialogue.

Crucially, a complex range of policy drivers and economic, sociocultural and environmental changes is now linked to the sustainable management of upland areas in the United Kingdom (Price et al. 2002; Bonn et al. 2009a). This reflects the changing and complex nature of environmental governance, with most upland land uses, particularly farming, influenced by an increasing number of actors and agencies. In particular, public policy decisions, and changes in environmental regulation and funding, can have significant impacts on land-management practices and decision-making (McCrone et al. 2008; Condliffe 2009). Currently, the wide range of regulatory requirements and incentives influencing Scotland's uplands derives from UK membership of the EU, the British government, and the devolved powers of the Scottish Parliament and its associated non-departmental government bodies (which often enforce EU directives translated into Scottish legislation). As a result, the uplands are increasingly contested, reflecting the interplay of power and influence in such settings and the increasing pressure for integrated delivery of multiple functions (Price and Holdgate 2002; Holden et al. 2007; Bonn et al. 2009a; Reed et al. 2009).

RECOGNISING ECOSYSTEM SERVICES WITH UPLAND MANAGERS IN MIND

The growing emphasis on recognising, understanding and operationalising the range of ecosystem services associated with uplands reflects this shift from sectoral to more integrated approaches. Research carried out in 2009 estimated the value of ecosystem services and natural capital in Scotland at 21.5 to 23 billion pounds (Williams 2009). As demands for services intensify, the provisioning roles of uplands are likely to become increasingly important, and landowners and managers will have crucial roles to play in the continued maintenance, enhancement and delivery of these services. The discussion below addresses a number of ecosystem services in the context of Scotland's uplands, making particular links to their relevance to the management of Scotland's upland estates in order to provide the context for the following chapters.

From 'traditional' to 'consumptive' enterprises: provisioning services

For at least 150 years, British uplands have been dominated by three 'traditional' land-use practices that are associated with provisioning services: sheep farming, forestry and game management (Sotherton et al. 2009). It should be noted that it is not always simple to disassociate the various services; thus game management is discussed below as a cultural service, given that its contribution to provisioning services, that is, the production of food, is of lesser importance. In many parts of Scotland's uplands, these 'traditional' enterprises are now 'struggling to produce significant income' (Lindsay and Thorp 2011, 417) and rely increasingly on subsidy payments. Provisioning services are strongly demand-led by external markets and policies, in particular, food security (Bonn et al. 2009b). As a result, they are frequently directly driven by intentional management that seeks to optimise their value (Zhang et al. 2007). Table 2.1 lists the main provisioning services associated with British uplands. We now consider

TABLE 2.1 Provisioning services in UK uplands (Adapted from: Bonn et al. 2009b)

Provisioning services	Upland examples
Food (livestock, crops)	Mainly used for livestock grazing by sheep and beef cattle. Low agricultural productivity, generally classed as poor agricultural land or 'Less Favoured Areas' (Baldock et al. 2002). Sporting estate management generates supplies of venison and game birds for sale to commercial game dealers (see Cultural services section also).
Fibre (timber, wool)	Sheep wool: a by-product with little agricultural value, though it also used as an insulating material. Forestry policies giving prominence to sustainable forest management and the contribution of woodlands to other ecosystem services (Mason 2007).
Minerals	The extraction of minerals for building stone, aggregate and lime has long been a feature of upland landscapes (Reed et al. 2009).
Energy provision	The topography and climate of the uplands mean that they are important current and potential sources of hydroelectric power and wind energy. Localised peat extraction is common (Tomlinson 2010): the impacts of its removal on biodiversity and carbon dioxide emissions are more widely recognised (DEFRA 2009). Wood fuel can be derived from forests.
Fresh water provision	Approximately 70 per cent of UK drinking water flows from upland catchments (Heal 2003). Land management and water quantity and quality are closely linked (e.g. grazing levels).

upland agriculture and forestry in more detail, as well as the growing emphasis placed on uplands as a place for renewable energy provision.

Upland agriculture: a less favoured enterprise?

Agriculture in Scotland's uplands is constrained by climate, soil and topography: heavy rainfall, acidic soils and little shelter give few options beyond hill sheep farming and some cattle grazing. Livestock farming has shaped and maintained semi-natural open moorland habitats (Bonn et al. 2009b), with both EU and British/Scottish policy acting as key drivers of change, particularly with respect to upland livestock numbers (McCrone et al. 2008; Condliffe 2009). Government financial incentives between 1945 and 1980, based on the need for increased food security, encouraged farmers to increase the productivity and fertility of upland soils through agricultural intensification (Evans et al. 2002; SAC 2008; Reed et al. 2009), generally using lime and fertilisers, drainage, ploughing and the conversion of native vegetation to more productive variants (Condliffe 2009). This led to the degradation of upland soils and habitats in many parts of Britain, with stocking densities increasing by up to 400 per cent in some areas as a result of subsidies paid according to the number of stock (Holden et al. 2007). This increase in stocking numbers was blamed for significant loss of heather moorland, the introduction of *Molinia* monocultures, and increases in bracken and gorse (Scott 1986).

Since 1980, with the introduction of quotas and stronger measures to protect the upland environment, there has been a gradual shift away from production-based subsidies (including livestock headage payments) towards incentives that support the increased provision of environmental goods and services. Specifically, reforms of the CAP – including the MacSharry (1992) and Agenda 2000 reforms – led to subsidy payments being fully decoupled from production in 2003 and increased the focus on environmental aspects through a greater emphasis on Pillar II of the CAP (McCrone et al. 2008; Condliffe 2009; Dupraz et al. 2011). These measures have led to a decline in sheep stocking numbers across many parts of the British uplands (SAC 2008; Commission for Rural Communities 2010). In Scotland, the total number of adult sheep decreased by more than 1.4 million from 1999 to 2009 (Scottish Government 2009). While such decreases may have negative impacts on species composition and vegetation structure, in turn affecting insect, bird, and mammal populations (for example, Holland et al. 2008), it is also likely that some species and habitats will benefit from the removal of livestock (Midgley and Price 2010). Hill farming also contributes to habitat and landscape maintenance, including maintaining traditional landscape features, such as hedges and drystone walls (Hubacek et al. 2009).

Farming remains a key element of upland estate businesses, though farming in the uplands for the sale of products alone is economically unsustainable, with farmers largely dependent on subsidies, in particular from agri-environment schemes and the Less Favoured Area classification (as outlined in Chapter 1), for economic survival. Land use therefore remains heavily influenced by the economic incentives deriving from EU and Scottish policy. Given the economically marginal nature of much upland farming, land managers in the uplands have increasingly had to seek non-agricultural sources of income (see Box 2.4). Such diversification or pluriactivity can involve charging for countryside services related to the land, such as farm tours and off-road vehicle courses, or converting land or property for other uses (for example, clay pigeon shooting, outdoor pursuits, bed and breakfast, renewable energy). Estates in prime tourist areas can generate significant income from such ventures. Often, farm incomes are also supplemented from off-farm part-time employment (examples can be found in Chapters 4 and 5). Opportunities for diversification are limited in many areas, however, particularly when those areas are not well served by infrastructure and/or are distant from major tourist routes; most farming families in the uplands are still largely dependent on agriculture for their income.

Towards sustainable forest management and renewable energy provision

After millennia of decreasing forest cover (Smout 2000), the twentieth century saw large-scale afforestation of upland areas, with the widespread conversion of semi-natural vegetation to plantation forests dominated by coniferous species (Reed et al. 2009). More recently, forestry policies have started to emphasise 'sustainable forest management' and the contribution of woodlands to ecosystem services (Mason 2007).

> **Box 2.4** Diversification and integrated management on the Glenlivet Estate (Sources: Wells 2004; Parmee 2006; Crown Estate 2012)
>
> The Glenlivet Estate covers 23,000 hectares on the edge of the Cairngorm Mountains of north-east Scotland and has been part of The Crown Estate since 1937. Nearly two-thirds (14,750 hectares) of the estate comprises farms, mainly leased to family businesses, with livestock as the main product. Sporting rights are also leased and include two major grouse moors, roe and red deer stalking, rough shooting, and a salmon beat. Farming and sporting activities both provide important sources of income for the estate. Forestry activities focus on producing a continuous crop of timber, and steps have also been taken to expand semi-natural woodland. The Glenlivet Estate Ranger Service is responsible for access management and interpretation; a large path network has been developed over the past decade. The long-term policy for the estate aims to ensure that the various land uses are managed in a fully integrated manner, providing sustained employment in land-based activities.

The Scottish Forestry Strategy (2006) aims for 25 per cent woodland cover nationally by the second half of the present century, with emphases on: increasing the proportion of native trees, both coniferous and deciduous; diversifying woodland structure; and improving the integration of forest management with other land uses, such as deer management (see Box 2.3).

From the late 1980s onwards, community groups have also become increasingly involved in forest and woodland management through direct community purchases and through the development of management agreements with Forestry Commission Scotland (FCS) and other stakeholders (Warren 2009). The community woodlands sector was boosted in 2003 by the formation of the Community Woodlands Association to represent this emergent movement nationally. The process of communities acquiring woodlands was also further facilitated in 2005 with the launch of the National Forest Land Scheme which provides communities with the opportunity to purchase or lease FCS land in situations where community ownership is viewed as likely to result in increased delivery of public benefit (FCS 2010). Currently, more than two hundred groups are involved in community woodlands across Scotland.

The Scottish Government's strong promotion of renewable energy offers landowners new opportunities, such as wind-power generation, hydroelectricity, and various sources of biomass energy. In particular, the re-emergence of wood, the most ancient of fuels, as a desirable, carbon-neutral energy source with the potential to contribute significantly to the mitigation of climate change may bring particular benefits to forest owners. As well as utilising the 'waste' products from forest management for energy, there is renewed interest in managing woodland specifically for the production of wood fuel via the historic method of short-rotation coppice or the more novel approach of short-rotation forestry. The potential for energy generation is considerable, especially for heat production (Andersen et al. 2005; Sustainable Development Commission 2005), though the inaccessibility and rough topography of many upland forests will make it challenging to realise that potential. Nevertheless, by enabling greater use of the timber resource, including low-value wood, expanding demands for wood fuel provide an incentive for woodland owners to manage undermaintained forests to a higher standard (Strachan and Beck 2008), and this has socioeconomic and environmental benefits. Indeed, Warren (2009, 91) comments that 'bioenergy seems to "tick all the boxes" as an exemplar of sustainable development'. Given that it is being actively encouraged at EU, UK and Scottish levels, it seems set to become an increasingly prominent component of upland estate enterprises.

Many landowners are also investing in run-of-river hydro schemes, dams and reservoirs, and wind power projects to take advantage of the generous feed-in tariff payments provided by the Scottish Government. On the one hand, the generation of hydroelectricity can provide income and be used to support local enterprises but, on the other, the construction of new dams and reservoirs may have significant impacts on river and riparian ecosystems (Warren 2009). The potential for generating wind energy is often limited by the infrastructure costs associated with transmitting energy from remote locations (Orr et al. 2008). Wind-energy infrastructure is also the subject of much debate within upland communities and wider communities of interest, signalling a tension between people's appreciation of perceived 'wild landscapes' and their desire to develop renewable energy sources and additional opportunities for economic development (this is discussed further in Chapter 6). For some estates with limited economic options, small-scale wind-turbine development can help transform the bottom line from red to black (Scott et al. 2008), and owners can also receive substantial ground-rent payments by 'hosting' large commercial wind farms on their land.

Experiencing uplands: cultural services

Many of Scotland's uplands are iconic landscapes of exceptional scenic beauty that inspire people and form part of Scotland's cultural and national identity, as reflected in their designation under European and national legislation (Lorimer 2000; Simmons 2003; SNH 2008). They offer 'attractive scenery, exposure to the elements in often remote locations, areas to engage in outdoor pursuits, sites of historic human artefacts and a wide variety of plant and animal species' (van der Wal et al. 2011, 115). Cultural services therefore include the enjoyment that people gain from upland landscapes, biodiversity and cultural heritage which have been the main drivers for their protection (see Table 2.2). The National Parks (Scotland) Act 2000 led to the establishment of two national parks that cover 7.1 per cent of Scotland (Reed et al. 2009); this was an important symbolic project for the devolved Scottish Parliament (Stockdale and Barker 2009). Protection of the natural heritage has also been a major incentive for the purchase of estates by national conservation organisations, and the topic of an increasing policy debate around the importance of landscapes and 'wild land' (see Chapter 7 for further discussion of wild land).

TABLE 2.2 Cultural services in UK uplands (Adapted from: Bonn et al. 2009b)

Cultural services	Upland examples
Recreation, tourism, education	Considerable increases in hillwalking in the twentieth century (Warren 2009; Hanley et al. 2002). Most visitors are attracted by the scenery and tranquillity (Davies 2006). Upland areas offer opportunities to learn about the natural and cultural heritage. Guided footpaths, visitor centres, school visits, etc. offer learning.
Field sport and game management	Grouse moor management is responsible (along with grazing) for shaping heather moorland habitats.
Landscape aesthetics (scenic beauty, sense of wildness)	Spiritual and religious reflection have been linked to travelling through wild and beautiful terrain and having uninterrupted views (Natural England 2009). Aesthetic features include remoteness, bleakness and open space – and the emerging concept of 'wild land' (Fisher et al. 2010).
Cultural heritage	Symbolic cultural assets and heritage can be found in the uplands (e.g. monuments, relics etc.) (Simmons 2003). Scotland's uplands have been promoted as symbols of national identity (Lorimer 2000) and include Scotland's two national parks. British uplands have worked as major inspiration for generations of writers, poets and artists, as well as providing cognitive and educational stimuli as dynamic, living landscapes (Reed et al. 2009).
Biodiversity	Maintenance and enhancement of biodiversity are intrinsically linked to human well-being and health benefits.
Health benefits	Upland areas provide physical and mental health benefits (Pretty et al. 2007), but are also potentially dangerous places.

Upland biodiversity: habitats that span landscapes

A significant proportion of Scotland's uplands is protected by conservation designations, both national and European. Under the latter, 12 per cent of the uplands is designated as Special Areas of Conservation (SACs) under the EU Habitats Directive, and 6 per cent as Special Protection Areas (SPAs) under the Birds Directive (Price et al. 2002).[2] These statistics show the importance of Scotland's uplands for preserving elements of biodiversity of significance at the European scale. At the habitat (SAC) level, these include heather moorland, one of Europe's most distinctive habitats; the remnants of the native Caledonian pinewood, an important western representative of Europe's boreal forest; and high mountains such as the Cairngorm plateau with important arctic–alpine habitats and permanent snow beds. Notably, 25 per cent of the Cairngorms National Park is designated as Sites of Special Scientific Interest (SSSI), and 49 per cent as Natura sites. These designations overlap and most are mainly on private land. The situation is similar in Scotland's other national park, Loch Lomond and the Trossachs, which is also in the uplands.

Upland habitats are home to many rare species: for example, a quarter of Britain's threatened animal, bird and plant species live in the Cairngorms National Park, both within and outside designated sites (Cairngorms National Park Authority 2013). Across the uplands, changes in habitats and species are being recorded (Midgely and Price 2010). Between 1998 and 2007, the condition and species richness of several habitats declined; alpine vegetation is becoming more homogeneous. Conversely, upland forests are increasing in extent. Similarly, there are contrasting trends among upland birds: from 1994 to 2007, curlew, black grouse, peregrine, and golden plover decreased in number while hen harrier, raven and snipe increased.

The conservation of biodiversity requires management not only in designated sites but also at the landscape scale: a key reason for the British government's adoption of the European Landscape Convention and the establishment of Scotland's two national parks and forty National Scenic Areas (NSAs).[3] The last designation recognises that certain landscapes have outstanding scenic value in a national context and therefore require special protection measures (SNH 2000). Thus, development within NSAs, most of which are in upland areas, is subject to more stringent planning procedures. The need for landscape-scale approaches is also recognised through the development and implementation of regional Biodiversity Action Plans which bring together a wide range of stakeholders to act jointly not only to maintain populations of species with intrinsic or economic (for example, game birds, deer, fish) value but also in recognition that careful management of terrestrial and aquatic habitats is important for providing a wide range of ecosystem services. This will become increasingly important as climate change interacts ever more with the effects of land management activities to influence all aspects of upland biodiversity.

Upland recreation: a growing industry

People are drawn to the uplands from other parts of Scotland, Britain, Europe, and further afield by the diverse landscapes and range of recreational opportunities (Price et al. 2002). Since the passing of the Land Reform (Scotland) Act 2003, the public has gained a formal right of 'responsible' access to all of the Scottish uplands (see Chapter 3 for more detail). Thus, uplands continue to be among the most popular tourist destinations, with 92.4 million visitor days per year to British upland national parks alone (Reed et al. 2009). In 2009, 2.2 million trips, adding up to almost 7 million visitor days, were taken in Loch Lomond and the Trossachs National Park (Loch Lomond and the Trossachs National Park Authority 2012), and 1.5 million people visited the Cairngorms National Park in 2011 (Global Tourism Solutions UK Ltd 2011). The economic importance of tourism to the upland economy in Britain was made clear during the foot-and-mouth disease outbreak in 2001 when access restrictions resulted in estimated losses to the tourism sector of up to £3.2 billion, while farmers received much less compensation – £1.34 billion – for livestock loss (Donaldson et al. 2006; Curry 2009). In 2011, tourism was estimated to contribute £4.5 billion to the Scottish economy (VisitScotland 2011). In the Cairngorms National Park, dependence on the primary sector has waned considerably and tourism has come to dominate, accounting for nearly 30 per cent of all gross value added (GVA) in the park, and employing 22 per cent of the workforce (Cairngorms National Park Authority 2011: 3).

Over the last few decades, there has been a large increase in outdoor recreation in Scotland's uplands, with hillwalking, mountain biking, rock climbing, mountaineering and skiing among the popular choices. Areas with concentrations of Munros [peaks over 3,000 feet (914 metres)], such as the Cairngorms and the West Highlands, have seen significant increases in the numbers of hillwalkers since the 1980s (Warren 2009), and the number of people who have climbed all of the Munros (Compleatists) over this period has increased comparably (Scottish Mountaineering Club 2012). Even remote Munros are now climbed by almost a thousand people a year, and the easily reached ones see many thousands of ascents annually. For example, Ben Lomond attracts more than 50,000 walkers a year (Loch Lomond and the Trossachs National Park Authority 2005). Downhill skiing generates £11 million for the Scottish economy (van der Wal et al. 2011) and, in the snow-free months, mountain biking is also very popular; Scotland is a 'global destination' for those who enjoy the sport (Scottish Mountain Bike Development Consortium 2009; Pothecary 2012). Visitors for whom mountain biking is a key reason for their visit contribute £46 million to Scotland's economy; the total number of mountain biking visits exceeds 1.3 million, half to purpose-built trail centres (Tourism Intelligence Scotland 2010). The 7stanes Mountain Biking initiative, which runs seven purpose-built trail centres across the Southern Uplands, reported net additional impacts of £9.29 million, 204.6 full-time equivalents (FTEs) and £3.72 million GVA in 2007 (EKOS Limited/Tourism Research Company 2007). The growing numbers of people accessing the uplands for recreation have led to increased erosion and disturbance to natural habitats, livestock, game and wildlife (McMorrow et al. 2009) though the extent of these impacts is subject to debate. Equally, purpose-built facilities, such as mountain biking trail networks, can be used to attract people to less sensitive locations, thus minimising the risk of damage to more vulnerable sites. A further means of minimising the environmental impacts of recreation is through the (re)construction of path networks; thus, access management has become a more explicit activity on some upland estates (see Box 2.5). In general, little direct economic benefit occurs to local landowners and managers through recreation and tourism, with the exception of land-management businesses that have diversified (as discussed earlier) (Bonn et al. 2009b).

Field sports and game management

Another important upland activity is field sports (including grouse, deer and fishing); as game is mainly managed for recreational field sports rather than for food supply, this is discussed here as a cultural service (Bonn et al. 2009b). Over 2 million hectares in Scotland (27 per cent of all privately owned land) are managed for game, including deer stalking (Irvine et al. 2009). Several million pounds per year of private investment support game management in Britain, particularly as one main motivation for owning a sporting estate is often the hunting experience, rather than income generation (Sotherton et al. 2009). Chapter 4 considers in more detail landowners' motivations for purchasing sporting estates. Large areas of uplands in Britain, in particular heather moorland, are managed by landowners for red grouse (Hubacek et al. 2009) and an estimated 47,000 people in the United Kingdom take part in grouse shooting (Natural England 2009). Approximately 450 grouse-shooting moors cover 16,763 square kilometres of Britain; the majority (296 sq. km) are in Scotland (Richards 2004). Though

---

Box 2.5 Managing access to Scotland's upland estates

The management of footpaths on upland estates is costly: each metre of constructed mountain footpath can range from £100 per metre to £200 per metre for a steep, stone-pitched path (NTS 2012). Well-known examples of path maintenance have also been carried out and paid for by charitable landowners: the major work carried out on the Schiehallion path (the seventh most popular final Munro) cost £817,000, funding for which was raised by the John Muir Trust (Howie 2008); and the National Trust for Scotland spent £350,000 on the Ben Lomond paths in the 1990s (Johnston 2000). In all of these cases, the maintenance of such large-scale capital investments adds further ongoing expense.

While some estates are able to afford the initial investment, only a small minority can offset these costs with access-related income. As a result, more public funds have become available, normally on a match-funding basis, for access-related expenditure, such as signage and footpath repairs. For example, the Cairngorms Outdoor Access Trust (COAT) works with a number of partners (including landowners) to deliver its Cairngorms Mountain Heritage Project which develops, maintains, interprets and promotes mountain paths with the national park (COAT 2011).

---

few return a profit (Sotherton et al. 2009), it has been estimated that, during 2009, grouse shooting supported 1,072 full-time equivalent employees in Scotland (Fraser of Allander Institute 2010). Nonetheless, there has been a gradual decline in the number of grouse shot over the past forty years (Aebischer and Baines 2008), and the appetite for traditional countryside recreation has decreased because of sociocultural shifts in recreational preferences (Natural England 2006; Suckall et al. 2009).

Game management 'often involves the preservation of natural or semi-natural habitats and the biodiversity they contain' (Irvine 2012, 2); grouse management, in particular, contributes to the maintenance of large areas of heather moorland, a cultural landscape of international importance (Lindsay and Thorp 2011; van der Wal et al. 2011). The recent National Ecosystem Assessment (van der Wal et al. 2011), however, recognised increasing concerns over: the visual impact of muirburn;[4] potential damage to peat structure and subsequent carbon losses through atmospheric exposure in more heavily burnt areas (Yallop and Clutterbuck 2009); and the impacts of associated long-term predator control (the killing of mammals and birds regarded as 'vermin') on some elements of biodiversity. Predator control takes place to protect agricultural and hunting interests and, in some cases, there have been conflicts because of the potential impacts of hunting on the status of species of conservation importance (Thirgood and Redpath 2008). While predator control may benefit ground-breeding moorland birds, such as lapwing and curlew (Fletcher et al. 2010), persecution has resulted in the local extermination of several birds of prey, including hen harriers, golden eagle and red kite (van der Wal et al. 2011). This situation 'indicates the importance of working towards a consensus, whereby the diversity of birds and mammals is enhanced while also accommodating cultural practices, such as grouse management, which are perceived to play a vital role in rural economies' (van der Wal et al. 2011: 118; see also Thirgood and Redpath 2008; Mc Morran 2009; White et al. 2009).

In Scotland, many upland estates are maintained for deer and obtain their primary income from deer stalking (stalking lets and venison sales) (Reed et al. 2009). This has led, in some cases, to changes in habitats because of unsustainably high numbers of deer, and conflicts with adjacent land uses, such as forestry. Despite increased culling, deer numbers have grown considerably over recent years, with a range of associated social, financial and environmental implications (van der Wal et al. 2011) (Box 2.6).

Box 2.6 Managing deer on Scotland's upland estates

Deer are iconic species in Scotland, and red deer, in particular, have long been associated with Scottish culture and identity, as well as being important components of upland ecosystems. Deer no longer have natural predators in Britain and, as wild deer populations naturally increase, these must be controlled by people to maintain the welfare and health of the animals and to limit the potential for overgrazing and other negative impacts (Hubacek et al. 2009). The policy context for deer management in Scotland is laid down in the Deer (Scotland) Act (1996) and the Wildlife and Natural Environment Act (2011). A number of strategies for deer management promote collaborative and integrated approaches to the sustainable management of deer in Scotland (DCS 2000; DCS et al. 2008).

Deer in Scotland are *res nullius* – they belong to no one until they are killed or captured; the right to shoot deer belongs to the owner of the land on which the deer are found, however. Since the 1800s, deer stalking for sport, predominantly on large 'sporting' estates, has become well established (Wightman 2004) and populations of Scotland's two native species (red and roe) have increased substantially in recent decades (Warren 2009). Deer are controlled not only on sporting estates but also by farmers and foresters because of their impacts on crops and young trees (Gill 1992; Putman et al. 1998; Gill 2000). There is also an increasing emphasis on reducing deer numbers, particularly on designated sites (Landwise Scotland 2005) because their occurrence at relatively high densities (thirty to forty deer per square kilometre or higher) on some sites has been associated with the degradation of valuable habitats through overgrazing and the inhibition of native woodland regeneration (Staines et al. 1995; Milne et al. 1998; Hunt 2003). It is important to note, however, that other herbivores, such as sheep, can also cause equal or greater damage than deer to upland habitats (Albon et al. 2007).

Deer represent an economic asset in terms of sporting incomes (and, to a lesser extent, venison sales); high numbers of deer, particularly stags, have long been equated with higher estate valuations (PACEC 2006; Campbell and Danson 2008; Warren 2009). Particularly in the Highlands, sporting estates provide employment, such as gamekeeping positions and general estate staff; deer management, and both private and commercial stalking activities supported 2,520 jobs in rural Scotland in 2005 and generated £105 million, two-thirds of which was retained in Scotland (Rose 2010).

Deer management faces the challenge of integrating management both on single landholdings – with deer often currently excluded from forestry using deer fences, for example, to protect young trees from browsing – and across estates, as deer move across ownership boundaries. Deer Management Groups (DMGs), comprising groups of neighbouring landowners and/or managers, have been established throughout the uplands with the aim of managing deer collaboratively at landscape scales (Davies and White 2012). While there are successful examples of DMGs, it is not clear how well they are currently resourced or work in practice (Nolan et al. 2002; Rose 2010). Nevertheless, DMGs represent the cornerstone of applying an integrated, ecosystem-management-based approach to deer management in Scotland, with GIS-based mapping techniques offering considerable potential for the future application of the DMG model (Davies and Irvine 2011).

The recent voluntary Code of Practice on Deer Management (SNH 2011), in conjunction with a suite of online best-practice guides (http://www.bestpracticeguides. org.uk), represent a further step forward in terms of moving towards sustainable outcomes for deer management.

WHAT ROLE FOR CLIMATE-CHANGE MITIGATION AND RISK MANAGEMENT?

Uplands serve an important function in global climate regulation, and there is increasing awareness that they provide a wide range of regulating services, including water purification, carbon storage and carbon sequestration, as well as potential for some flood regulation (Bonn et al. 2009b; van der Wal et al. 2011). UK peatlands – including upland blanket bogs, lowland raised bogs and fen peats – contain 5.1 billion tonnes of carbon, of which the majority, 4.5 billion tonnes, is in Scotland (Ostle et al. 2009). It is only relatively recently that attention has been drawn to the potential of the uplands for carbon storage, which may become economically significant owing to carbon pricing policies aimed at climate-change mitigation (Hubacek et al. 2009). Over £150 million has been spent on peatland restoration in the British uplands since 2005 (Reed et al. 2009). Within the British carbon inventory, peatlands are considered a net sink (Worrall et al. 2003), and much of Scotland's peatlands are within upland estates (Reed et al. 2011), giving estate managers a responsibility in terms of carbon storage and climate-change mitigation, along with the management of other regulating/supporting services (see Table 2.3).

Much of Britain's peatland is degraded, however, with erosion and overburning leading to significant exposure. A range of new planning tools, such as the community infrastructure levy, biodiversity offsetting and visitor payback schemes, may collectively contribute to improved management of these peatlands. This requires improved professional and public understanding and appreciation of the wider value of peatlands as carbon stores, however, as well as the other ecosystem services they provide (Reed et al. 2011).

TABLE 2.3 Regulating and supporting services in UK uplands (Adapted from: Bonn et al. 2009b)

Regulating services	Upland examples
Climate regulation: carbon storage and sequestration	Changes in climate, notably in rainfall and temperature, are likely to have an impact on the capacity of ecosystems to store carbon. Preventing soil erosion and enhancing peat decomposition reduce impacts on water quality, loss of water storage capacity, and carbon loss.
Air quality regulation	Uplands are main areas for atmospheric deposition of pollutants due to higher levels of precipitation and cloud deposition, thereby 'cleaning' the air (Caporn and Emmett 2009).
Water quality regulation	Plant–soil systems of upland habitats intercept and retain a proportion of several atmospheric pollutants.
Flood risk prevention	Limited evidence whether the UK's upland habitats act to attenuate or exacerbate flooding (Orr et al. 2008).
Wildfire risk prevention	Wildfires are one of the top twenty-five priority risks to UK biodiversity (McMorrow et al. 2009).
Supporting services	
Nutrient cycling, water cycling, soil formation	Support for all other services listed in this chapter.
Habitat provision for wildlife	Most key ecosystem services are underpinned by ecosystem processes and habitat structure in the uplands. Degradation, conversion and fragmentation threaten the provision of wildlife habitats.

UPLAND COMMUNITIES AND ECONOMIES

Scotland's uplands are the most sparsely settled parts of the country and have some of the lowest population densities in Europe (Price et al. 2002). The shift towards a 'post-productivist' era (as outlined in Chapter 1) has significantly altered rural economies, including a decline in forestry, fishing and agricultural employment opportunities, with most employment in the service sector (Ward 2006). Rural wages are generally lower than in urban areas (Hall and Skerratt 2010) and there is a lack of available affordable housing, owing to rising house and land prices, increasing household numbers, and the purchase of both main and second homes by people with high incomes (this is discussed further in Chapter 5) (Skerratt et al. 2012a). Quality of life and environment are increasingly important drivers for business location, and significant progress has been made with rural diversification through the conversion of derelict farm buildings to industrial units, enabling the revitalisation of certain areas subject to positive planning policies (Scott et al. 2006). Improvements in communications technology have also boosted opportunities for businesses to locate in rural areas (Ward 2006) but there is still limited access to high-speed broadband Internet in many remote rural areas, and this works as a barrier to such developments (Skerratt et al. 2012b).

Many communities in upland Scotland have an ageing population, and emigration of young people is common. This is combined with a continuing strong current of urban to rural migration which has resulted in the reconstruction of many rural communities (Stead 2000; Woods 2005), creating a demographic profile skewed towards those who are no longer economically active (though there are variations in demographic trends between accessible rural and remote rural areas; see Thomson 2010). So-called 'incomers' come from a variety of backgrounds, age groups, occupations and household income levels, though the mobile and middle-class tend to predominate (Woods 2005; Hall and Skerratt 2010). Such 'amenity migrants' (Moss 2006) can bring ideas, experience, and capital which can be of great benefit in developing local economies but, at the same time, they can also create tensions because of new ways of thinking and of doing things, and by driving up property prices. This is exacerbated in popular tourist areas where second home ownership can threaten community cohesion and resilience. Frequently, inflated house prices in rural areas affect young people and the elderly who struggle to remain in their home area, close to family (Jones 2001). The rise in second home ownership also results in issues of 'out-pricing' (Rye 2011); remote rural areas have the highest proportions of second home ownership in Scotland (Skerratt et al. 2012b). This leads to communities becoming socially polarised and losing key services as well as community identity.

A key mechanism which tries to address all these competing demands is the planning system which translates policies from the national level through Scottish planning policy notes through to development plans that set out the spatial pattern of growth over the next twenty years. Significantly, there has been a marked shift away from protecting the countryside for its own sake in favour of a more sustainable growth ethic that encourages rural revitalisation. The planning system itself is still ill-prepared to deal with rural development problems, however, given the way that urban and rural planning systems have, in effect, been separated (Curry 1993; Scott et al. in press). The focus on protecting rural landscapes from development has led to significant constraints on housing supply outside conventional urban centres, thus increasing the price of development land which, in turn, leads to high prices for new housing, and therefore difficulties in meeting demands for affordable housing (Home 2009). Chapter 5 considers these issues in more detail, particularly focusing on the role of the private landowner in developing employment and housing opportunities for local communities.

WHAT ARE THE LESSONS FOR SUSTAINABLE GOVERNANCE?

The uplands provide multifunctional assets that benefit society in many different ways. Fundamentally, the continued delivery of provisioning, cultural, regulating and supporting ecosystem services requires well-functioning, extensive and intact ecosystems; losses in scale and deteriorations in quality constrain the supply of ecosystem services (van der Wal et al. 2011). To improve decisions on the use and management of the uplands, the ecosystem services framework provides a useful lens within which it is possible to start to capture these values and understand how policy decisions are likely to affect ecosystem services in the future (Reed et al. 2013).

Decision-making in Scotland's uplands is driven by a combination of economic, environmental, social and political factors, including the motivations and aspirations of landowners and managers. The key driver, however, is currently the Scottish Government's policy of sustainable economic growth: a specific focus that may be in contradiction to the more integrated frameworks to plan for, and manage, the uplands, as proposed in this chapter and also espoused, for instance, in the Land Use Strategy. At the same time, decisions that were once dictated mainly by individual and local considerations are now subject to a rapidly expanding and multiscalar governance agenda (as introduced in Chapter 1) influenced by an array of management, agency and stakeholder perspectives, and captured within the different ecosystem services discussed in this chapter. Despite the perception that landowners retain 'exclusive dominion' (Hurley et al. 2002), their freedom of manoeuvre is now significantly constrained, and their rights exist alongside the powers and rights of both the state and the public. The tensions and conflicts which result from this uneasy coexistence are explored in more detail in the next chapter.

The management of Scotland's uplands, and those of Britain as a whole, exists within a context of diverse private and public preferences and uncertainty about future financial support. In the face of declining and uncertain incomes, particularly from farming, landowners and managers are increasingly diversifying into new ventures – for instance, in conservation, renewable energy, recreation and tourism – in the search for additional income. Several of these have been described in this chapter. There is, therefore, a need to understand better the priorities and management objectives of those who own and manage land in the uplands, recognising that, while they provide income and employment, the uplands are also cultural landscapes with multiple functions and benefits to wider society (Sotherton et al. 2009). Upland areas 'must continue to provide the many functions and services we have come to expect from them' without compromising biodiversity and important landscape features (Reed et al. 2009, S205), and this requires the integration of various objectives and the involvement of key players.

Multiple agencies and actors have stakes in how the uplands are managed, particularly as management for the production of sheep, grouse, deer, or trees may not necessarily be compatible with the provision of other environmental and cultural services (Hubacek et al. 2009; Curry 2009; Quinn et al. 2010). It is therefore increasingly acknowledged that it is necessary to understand the sustainability perceptions of the wide range of upland stakeholders and 'explore the range of emerging approaches to governance that better align community interests and rural planning' (Reed et al. 2009, S213; Swales 2009). While inequalities between stakeholders and conflicting objectives can function as barriers to effective land management for the public good (Quinn et al. 2007), there is a growing need to develop improved decision-making and practice for upland environments, incorporating the generation of collaborative knowledge and solutions for sustainability. The following chapters provide more in-depth assessments of the diverse approaches to management and the values of landowners and managers in achieving their goals, with regard to experiences, actions and constraints on a number of upland estates.

NOTES

1. For more detail on many of the topics in this chapter, see 'Managing Scotland's Environment' (Warren 2009), the 'Mountains, Moorlands and Heaths' chapter of the UK National Ecosystem Assessment (van der Wal et al. 2011) or other chapters in this volume, as signposted.
2. Together, SACs and SPAs are referred to as Natura sites.
3. National Scenic Areas are recognised by the International Union for the Conservation of Nature (IUCN) as Category V Protected Landscapes.
4. In Scotland, the Muirburn Code was amended as an outcome of the Wildlife and Natural Environment Act (2011), setting out regulations for when and how muirburn can take place (Scottish Government 2011d).

REFERENCES

Aebischer, N. J. and Baines, D. (2008). 'Monitoring gamebird abundance and productivity in the UK: The GWCT long-term datasets'. *Revista Catalana d'Ornitologia* 24, pp. 30–43.

Albon, S. D., Brewer, M. J., O'Brien, S., Nolan, A. J. and Cope, D. (2007). 'Quantifying the grazing impacts associated with different herbivores on rangelands'. *Journal of Applied Ecology* 44, pp. 1176–87.

Allmendinger, P. (2003). 'Integrated Planning in a Devolved Scotland'. *Planning Research and Practice* 18 (1), pp. 19–36.

Andersen, R. S., Towers, W. and Smith, P. (2005). 'Assessing the potential for biomass energy to contribute to Scotland's renewable energy needs'. *Biomass and Bioenergy* 29 (2), pp. 73–82.

Baldock, D., Dwyer, J., Peterson, J.-E. and Ward, N. (2002). *The Nature of Rural Development: Towards a Sustainable Integrated Rural Policy in Europe. A Ten Nation Scoping Study Synthesis Report.* Institute for European Environmental Policy, London.

Bonn, A., Allott, T., Hubacek, K. and Stewart, J. (2009a). 'Introduction: Drivers of Change in upland environments: concepts, threats and opportunities'. In: Bonn, A., Allott, T., Hubacek, K. and Stewart, J. (eds), *Drivers of Environmental Change in Uplands.* Routledge, London and New York, pp. 1–10.

Bonn, A., Allott, T., Hubacek, K. and Stewart, J. (2009b). 'Conclusions: Managing change in the uplands – challenges in shaping the future'. In: Bonn, A., Allott, T., Hubacek, K. and Stewart, J. (eds), *Drivers of Environmental Change in Uplands.* Routledge, London and New York, pp. 475–94.

Brooker, R. (2011). 'The Changing Nature of Scotland's Uplands – an interplay of processes and timescales'. In: Marrs, S. J., Foster, S., Hendrie, C., Mackey, E. C. and Thompson, D. B. A. (eds), *The Changing Nature of Scotland.* TSO Edinburgh, pp. 381–96.

Cairngorms National Park Authority (2011). 'Cairngorms National Park Strategy and Action Plan for Sustainable Tourism 2011–2016'. <http://cairngorms.co.uk/resource/docs/publications/06092012/CNPA.Paper.1840.Strategy%20and%20Action%20Plan%20for%20Sustainable%20Tourism%202011-2016.pdf> (last accessed 8 January 2013).

Cairngorms National Park Authority (2013). 'Cairngorms Nature Action Plan'. CNPA, Grantown-on-Spey.

Cairngorms Outdoor Access Trust (2011). 'COAT Annual Report 2010/2011'. <http://www.cairngormsoutdooraccess.org.uk/images/PDF/annualreport2011final.pdf> (last accessed 19 December 2012).

Campbell, I. and Danson, M. (2008). 'A Study to Scope the Key Social and Economic Impacts Associated with Large-Scale Reductions in Red Deer Populations'. Deer Commission for Scotland, Inverness.

Caporn, S. J. M. and Emmett, B. A. (2009). 'Threats from air pollution and climate change to upland systems: past, present and future'. In: Bonn, A., Allott, T., Hubacek, K. and Stewart, J. (eds), *Drivers of Environmental Change in Uplands*. Routledge, London and New York, pp. 34–58.

Commission for Rural Communities (2010). 'High Ground, High Potential. A Future for England's uplands communities'. Summary report. Commission for Rural Communities, Gloucester.

Condliffe, I. (2009). 'Policy change in the uplands'. In: Bonn, A., Allott, T., Hubacek, K. and Stewart, J. (eds), *Drivers of Environmental Change in Uplands*. Routledge, London and New York, pp. 59–90.

Countryside Commission for Scotland (1990). *The Mountain Areas of Scotland: Conservation and Management*. CCS, Battleby. 64 pp.

Crofts, R. (2000). 'Sustainable development and environment: delivering benefits globally, nationally and locally'. Scottish Natural Heritage Occasional Papers, No. 8.

Crown Estate (2012). 'Estate management'. <http://www.glenlivetestate.co.uk/estate_manage ment.html> (last accessed 20 December 2012).

Curry, N. R. (1993). 'Countryside planning: A look back in anguish'. Unpublished Inaugural Lecture, Cheltenham and Gloucester College of Higher Education, Cheltenham.

Curry, N. R. (2009). 'Leisure in the Landscape: Rural Incomes and Public Benefits'. In: Bonn, A., Allott, T., Hubacek, K. and Stewart, J. (eds), *Drivers of Environmental Change in Uplands*. Routledge, London and New York, pp. 227–90.

Davidson, D. A. (1994). 'Conservation as a land use in Scotland'. In: Fenton, A. and Gillmor, D. A. (eds), *Rural land use on the Atlantic periphery of Europe: Scotland and Ireland*. Royal Irish Academy, Dublin, pp. 173–84.

Davies, A. and Irvine, J. (2011). 'Mapping tools for collaborative deer management – A report for Cairngorms Speyside Deer management Group'. The James Hutton Institute and the outputs of the RELU project on DMGS <http://www.macaulay.ac.uk/RELU/dg_poster.php> (last accessed 8 January 2013).

Davies, A. L. and White, R. M. (2012). 'Collaboration in natural resource governance: reconciling stakeholder expectations in deer management in Scotland'. *Journal of Environmental Management* 112, pp. 160–9.

Davies, S. (2006). 'Recreation and visitor attitudes in the Peak District moorlands'. Moors for the Future Partnership, Rep. No. 12, Edale.

DCS (2000). *Wild Deer in Scotland: A long term vision*. Deer Commission for Scotland, Inverness. 12 pp.

DCS, FCS and SNH (2008). 'A Joint Agency Strategy for Wild Deer in Scotland'. DCS, Inverness.

DEFRA (2009). 'Safeguarding our soils. A Strategy for England'. Department of Environment, Food and Rural Affairs, London. <http://www.defra.gov.uk/environment/quality/land/soil/ documents/soil-strategy.pdf (last accessed 20 November 2011).

Dodgshon, R. A. and Olsson, G. A. (2006). 'Heather moorland in the Scottish Highlands: the history of a cultural landscape, 1600–1880'. *Journal of Historical Geography* 32, pp. 21–37.

Donaldson, A., Lee, R., Ward, N., Wilkinson, K. (2006). 'Foot and mouth – five years on: the legacy of the 2001 foot and mouth disease crisis for farming and the British countryside'. Centre for Rural Economy, University of Newcastle-upon-Tyne.

Dougill, A. J., Fraser, E. D. G., Holden, J., Hubacek, K., Prell, C., Reed, M. S., Stagl, S. and Stringer, L. C. (2006). 'Learning from doing participatory rural research: lessons from the Peak District National Park'. *Journal of Agricultural Economics* 57, pp. 259–75.

Dupraz, P., van den Brink, A. and Latacz-Lohmann, U. (2011). 'Nature preservation and production'. In: Oskam, A., Meester, G. and Silvis, H. (eds), *EU policy for agriculture, food and rural areas*. Wageningen Academic Publishers, The Netherlands, pp. 359–70.

EKOS Limited and Tourism Research Company (2007). '7 stanes Phase 2 Evaluation'. Report for Forestry Commission Scotland. <http://www.imba.com/sites/default/files/7StanesPhase2FinalReport.pdf> (last accessed 8 January 2013).

Evans, N., Morris, C. and Winter, M. (2002). 'Conceptualizing agriculture: a critique of post-productivism as the new orthodoxy'. *Progress in Human Geography* 26, pp. 313–32.

Forestry Commission Scotland (2010). 'National Forest Land Scheme guidance'. Forestry Commission Scotland, Edinburgh.

Fisher, M., Carver, S., Kun, Z., Mc Morran, R., Arrell, K. and Mitchell, G. (2010). 'Review of status and conservation of wild land in Europe. Project commissioned by the Scottish Government.

Fletcher, K., Aebischer, N. J., Baines, D., Foster, R. and Hoodless, A. N. (2010). 'Changes in breeding success and abundance of ground-nesting moorland birds in relation to the experimental deployment of legal predator control'. *Journal of Applied Ecology* 47 (2), pp. 263–72.

Fraser of Allander Institute (2010). *An economic study of Scottish grouse moors: An update (2010).* Game and Wildlife Conservation Trust, Perth.

Gill, R. M. A. (1992). 'A review of damage by mammals in North temperate Forests: 3. Impacts on trees and forests'. *Forestry* 65 (2), pp. 363–88.

Gill, R. M. A. (2000). 'The impact of deer on woodland biodiversity. Forestry Commission Information Note 36'. Forestry Commission, Edinburgh. 6 pp.

Global Tourism Solutions UK Ltd (2011). 'Cairngorms National Park STEAM Report 2011 – Badenoch and Strathspey Area, Rest of the Park'. 19 December 2012.

Hall, C. and Skerratt, S. (2010). 'What is the future for Scotland's rural infrastructure and access to services?' In: Skerratt, S., Hall, C., Lamprinopoulou, C., McCracken, D., Midgley, A., Price, M., Renwick, A., Revoredo, C., Thomson, S., Williams, F. and Wreford, A., *Rural Scotland in Focus 2010.* Rural Policy Centre, Scottish Agricultural College, Edinburgh, pp. 30–41.

Hanley, N., Alvarez-Farizo, B. and Shaw, W. D. (2002). 'Rationing an open-access resource: mountaineering in Scotland'. *Land Use Policy* 19, pp. 167–76.

Heal, W. (2003). 'Introduction, context, and seminar conclusions'. In: 'Managing Upland Catchments: Priorities for Water and Habitat Conservation'. Joint Nature Conservation Committee, Durham.

Holden, J., Shotbolt, L., Bonn, A., Burt, T. P., Chapman, P. J., Dougill, A. J., Fraser, E. D. G., Hubacek, K., Irvine, B., Kirkby, M. J., Reed, M. S., Prell, C., Stagl, S., Stringer, L. C., Turner, A. and Worrall, F. (2007). 'Environmental change in moorland landscapes'. *Earth-Science Reviews* 82, pp. 75–100.

Holland, J. P., Pollock, M. L. and Waterhouse, A. (2008). 'From over-grazing to under-grazing: are we going from one extreme to another?' *Annals of Applied Biology* 85, pp. 25–30.

Home, R. (2009) 'Land ownership in the United Kingdom: Trends, preferences and future challenges'. *Land Use Policy* 26 (1), S103–S108.

Howie, R. (2008). 'Schiehallion . . . and a too-easy ascent'. *The Scotsman,* 19 January. <http://www.scotsman.com/news/schiehallion-and-a-too-easy-ascent-1-1074357> (last accessed 8 January 2013).

Hubacek, K., Beharry, N., Bonn, A., Burt, T., Holden, J., Ravera, F., Reed, M., Stringer, L. and Tarrasón, D. (2009). 'Ecosystem Services in Dynamic and Contested Landscapes: The Case of UK uplands'. In: Winter, M. and Lobley, M. (eds), *What is the Land For? The Food, Fuel and Climate Change Debate.* Routledge, London, pp. 167–86.

Hunt, J. F. (2003). 'Impacts of wild deer in Scotland – How fares the public interest?' WWF Scotland and RSPB Scotland, Aberfeldy.

Hurley, J. M., Ginger, C. and Capen, D. E. (2002). 'Property concepts, ecological thought, and ecosystem management: a case of conservation policy making in Vermont'. *Society and Natural Resources* 15, pp. 295–312.

International Centre for Protected Landscapes (2008). 'Identifying Good Practice from countries implementing the European Landscape Convention'. Project Ref: ICP/001/07, International Centre for Protected Landscapes, Aberystwyth.

Irvine, J. (2012). "HUNTing for Sustainability' – A summary of research findings from the Scottish case study'. James Hutton Institute, University of Aberdeen and University of Stirling. <http://fp7hunt.net/Portals/HUNT/Reports/hunt%20scottish%20summary%20 july12.pdf> (last accessed 8 January 2013).

Irvine, R. J., Fiorini, S., McLeod, J., Turner, A., van der Wal, R., Armstrong, H., Yearley, S., and White, P. C. L. (2009). 'Can managers inform models? Integrating local knowledge into models of red deer habitat use'. *Journal of Applied Ecology* 46, pp. 344–52.

Johnston, J. L. (2000). *Scotland's Nature in Trust: the National Trust for Scotland and its wildland and crofting management.* Academic Press, London. 266 pp.

Jones, G. (2001). 'Fitting Homes? Young People's Housing and Household Strategies in Rural Scotland'. *Journal of Youth Studies* 4 (1), pp. 41–62.

Jordan, E. and Christie, M. (2010). 'Natural Heritage Futures: A Strong Vision for the Natural Heritage'. In: Marrs, S. J., Foster, S., Hendrie, C., Mackey, E. C. and Thompson, D. B. A. (eds), *The Changing Nature of Scotland.* The Stationery Office, Edinburgh, pp. 73–82.

Jordan, G. and Halpin, D. (2006). 'The political costs of policy coherence: constructing a rural policy for Scotland'. *Journal of Public Policy* 26 (1), pp. 21–41.

Kerr, G. (2004). 'The contribution and socio-economic role of Scottish estates: summary report'. Scottish Agricultural College, Penicuik.

Landwise Scotland (2005). 'Developing a policy framework for managing diffuse deer impacts'. Commissioned report for the Deer Commission for Scotland. Project RP35C.

Lindsay, Earl of and Thorp, S. (2011). 'Prospects for the Future: the Uplands in Peril or Thriving?' In: Marrs, S. J., Foster, S., Hendrie, C., Mackey, E. C. and Thompson, D. B. A. (eds), *The Changing Nature of Scotland.* The Stationery Office, Edinburgh, pp. 415–24.

Loch Lomond and the Trossachs National Park Authority (2005). 'State of the Park Report 2005: Land Based Recreation'. Loch Lomond and the Trossachs National Park Authority, Callander.

Loch Lomond and the Trossachs National Park Authority (2012). 'Annual Report and Accounts 2011–2012'. <http://www.lochlomond-trossachs.org/images/stories/Looking%20After/PDF/ publication%20pdfs/Annual_report_2011-12_web.pdf> (last accessed 8 January 2013).

Lorimer, H. (2000). 'Guns, Game and the Grandee: The Cultural Politics of Deerstalking in the Scottish Highlands'. *Cultural Geographies* 7 (4), pp. 403–31.

Lowe, P., Cox, G., MacEwen, M., O'Riordan, T. and Winter, M. (1986). *Countryside Conflicts: the politics of farming, forestry and conservation.* Gower Publishing, Aldershot. 378 pp.

McCrone, G., Maxwell, J., Crofts, R., Barbour, A., Kelly, B., Linklater, K., Ratter, D., Reid, D. and Slee, B. (2008). 'Committee of Inquiry into the Future of Scotland's Hills and Islands'. Royal Society of Edinburgh, Edinburgh.

Mc Morran, R. (2009). 'Red grouse and the Tomintoul and Strathdon communities – The benefits and impacts of the grouse shooting industry from the rural community perspective'. The Scottish Countryside Alliance Educational Trust Commissioned Report.

McMorrow, J., Lindley, S., Aylen, J., Cavan, G., Albertson, K. and Boys, D. (2009). 'Moorland wildfire risk, visitors and climate change: patterns, prevention and policy'. In: Bonn, A., Allott, T., Hubacek, K. and Stewart, J. (eds), *Drivers of Environmental Change in Uplands.* Routledge, London and New York, pp. 448–74.

Mason, W. L. (2007). 'Changes in the management of British forests between 1945 and 2000 and possible future trends'. *Ibis* 149, pp. 41–52.

Mather, A. S., Hill, G. and Nijnik, M. (2006). 'Post-productivism and rural land use: cul de sac or challenge for theorization?' *Journal of Rural Studies* 22, pp. 441–55.

Midgley, A., Williams, F., Slee, B. and Renwick, A. (2008). 'Primary Land-Based Business Study'. Scottish Agricultural College, Edinburgh.

Midgley, A. and Price, A. (2010). 'What future for upland biodiversity?' In: Skerratt, S., Hall, C., Lamprinopoulou, C., McCracken, D., Midgley, A., Price, M., Renwick, A., Revoredo, C., Thomson, S., Williams, F. and Wreford, A., *Rural Scotland in Focus 2010*. Rural Policy Centre, Scottish Agricultural College, Edinburgh, pp. 80–7.

Milne, J. A., Birch, C. P. D., Hester, A. J., Armstrong, H. M. and Robertson, A. (1998). 'The impact of vertebrate herbivores on the natural heritage of the Scottish uplands'. *Scottish Natural Heritage Review* 95. 127 pp.

Morris, A. and Robinson, G. (1996). 'Rural Scotland: problems and prospects'. *Scottish Geographical Magazine* 112 (2), pp. 66–9.

Moss, L. A. G. (2006). 'The amenity migrants: ecological challenge to contemporary Shangri-La'. In: Loss, L. A. G. (ed.), *The Amenity Migrants: Seeking and Sustaining Mountains and their Cultures*. CABI, Wallingford, pp. 3–25.

Mowle, A. (1997). 'The managing of the land'. In: Magnusson, M. and White, G. (eds), *The Nature of Scotland: Landscape, Wildlife and People*. Canongate, Edinburgh, pp. 133–43.

National Trust for Scotland (2012). 'The Footpath Fund – conserving the mountains and wild land you love'. <http://www.nts.org.uk/footpathfund> (last accessed 8 January 2013).

Natural England (2006). 'England Leisure Visits. Report of the 2005 Survey'. Natural England. Peterborough.

Natural England (2009). 'Experiencing landscapes: capturing the cultural services and experiential qualities of landscape'. Natural England Commissioned Report NECR024, Peterborough.

Nolan, A. J., Hewison, R. L. and Maxwell, T. J. (2002). 'Deer Management Groups: Operation and Good Practice'. Deer Commission for Scotland commissioned report. The Macaulay Institute, Craigiebuckler.

Orr, H. G., Wilby, R. L., McKenzie Heder, M. and Brown, I. (2008). 'Climate change in the uplands: a UK perspective on safeguarding regulatory ecosystem services'. *Climate Research* 37, pp. 77–98.

Ostle, N. J., Levy, P. E., Evans, C. D. and Smith, P. (2009). 'UK land use and soil carbon sequestration'. *Land Use Policy* 26 (S1), S274–S283.

PACEC (2006). 'The Contribution of Deer Management to the Scottish Economy'. Public and Corporate Economic Consultants, Cambridge.

Parmee, J. H. (2006). *An Evaluation of Community Engagement practices on the Glenlivet Estate, Scotland*. Unpublished MSc thesis, University of Edinburgh.

Pothecary, F. (2012). *What does responsible access in the uplands mean conceptually and in practice for mountain bikers and land managers in the Cairngorms National Park?* Unpublished MSc thesis, University of the Highlands and Islands.

Prager, K., Reed, M. S. and Scott, A. J. (2012). 'Viewpoint: Encouraging collaboration for the provision of ecosystem services at a landscape scale – Rethinking agri-environmental payments'. *Land Use Policy* 29, pp. 244–9.

Pretty, J., Peacock, J., Hine, R., Sellens, M., South, N. and Griffin, M. (2007). 'Green exercise in the UK countryside: Effects on health and psychological well-being, and implications for policy and planning'. *Journal of Environmental Planning and Management* 50, pp. 211–31.

Price, M. F., Dixon, B. J., Warren, C. R. and Macpherson, A. R. (2002). *Scotland's Mountains: Key Issues for their Future Management*. Scottish Natural Heritage, Battleby. 90 pp.

Price, M. F. and Holdgate, M. (2002). 'Sustainable Futures for the British Uplands'. Royal Geographical Society (with the IBG), London.

Putman, R. J. and Moore, N. P. (1998). 'Impact of deer in lowland Britain on agriculture, forestry and conservation habitats'. *Mammal Review* 28 (4), pp. 141–64.

Quinn, C. H., Huby, M., Kiwasila, H. and Lovett, J. C. (2007). 'Design principles and common pool resource management: an institutional approach to evaluating community management in semi-arid Tanzania'. *Journal of Environmental Management* 84, pp. 100–13.

Quinn, C. H., Fraser, E. D. G., Hubacek, K. and Reed, M. S. (2010). 'Property rights in UK uplands and the implications for policy and management'. *Ecological Economics* 69, pp. 1355–63.

Ratcliffe, D. A. and Thompson, D. B. A. (1988). 'The British uplands: their ecological character and international significance'. In: Usher, M. B. and Thompson, D.B.A. (eds), *Ecological Change in the Uplands*. Blackwell, Oxford, pp. 9–36.

Reed, M. S., Bonn, A., Slee, W., Beharry-Borg, N., Birch, J., Brown, I., Burt, T. P., Chapman, D., Chapman, P. J., Clay, G., Cornell, S. J., Fraser, E. D. G., Glass, J. H., Holden, J., Hodgson, J. A., Hubacek, K., Irvine, B., Jin, N., Kirkby, M. J., Kunin, W. E., Moore, O., Moseley, D., Prell, C., Price, M. F., Quinn, C., Redpath, S., Reid, C., Stagl, S., Stringer, L. C., Termansen, M., Thorp, S., Towers, W., Worrall, F. (2009). 'The future of the uplands'. *Land Use Policy* 26S, S204–S216.

Reed, M. S., Buckmaster, S., Moxey, A., Keenleyside, C., Fazey, I., Scott, A., Thomson, K., Thorp, S., Anderson, R., Bateman, I., Bryce, I., Christie, M., Glass, J., Hubacek, K., Quinn, C. H., Maffey, G., Midgely, A., Robinson, G., Stringer, L. C., Lowe, P. and Slee, W. (2011). 'Policy Measures for Sustainable Management of UK Peatlands'. IUCN Technical Review No. 8. UK IUCN Peatland Programme.

Reed, M. S., Hubacek, K., Bonn, A., Burt, T. P., Holden, J., Stringer, L. C., Beharry-Borg, N., Buckmaster, S., Chapman, D., Chapman, P. J., Clay, G. D., Cornell, S. J., Dougill, A. J., Evely, A. C., Fraser, E. D. G., Jin, N., Irvine, B. J., Kirkby, M. J., Kunin, W. E., Prell, C., Quinn, C. H., Slee, B., Stagl, S., Termansen, M., Thorp, S. and Worrall, F. (2013). 'Anticipating and Managing Future Trade-offs and Complementarities between Ecosystem Services'. *Ecology and Society* 18 (1), p. 5.

Richards, C. (2004). 'Grouse shooting and its landscape: The management of grouse moors in Britain'. *Anthropology Today* 20, pp. 10–15.

Rye, J. F. (2011). 'Conflicts and contestations. Rural populations' perspectives on the second homes phenomenon'. *Journal of Rural Studies* 27 (3), pp. 263–74.

SAC (2008). 'Farming's retreat from the hills'. Scottish Agricultural College Rural Policy Centre, Edinburgh.

Scott, A. J. (1986). Iss*ues in Common Land Management: A Case Study of the Dartmoor Commons*. Unpublished PhD thesis, University of Wales, Aberystwyth.

Scott, A. J. (2011). 'Beyond the conventional: Meeting the challenges of landscape governance within the European Landscape Convention?' *Journal of Environmental Management* 92, pp. 2754–62.

Scott A. J., Midmore, P. and Christie, M. (2004). 'Foot and Mouth: implications for rural restructuring in Wales'. *Journal of Rural Studies* 20, pp. 1–14.

Scott, A. J., Shannon, P. and White, P. (2006). 'Planning for Rural Diversification in Aberdeenshire'. Report by University of Aberdeen and the Macaulay Institute, Aberdeen.

Scott, A., Gilbert, A. and Gelan, A. (2007). 'The Urban–Rural Divide: Myth or Reality?' SERG Policy Brief Number 2. Macaulay Institute, Aberdeen.

Scott, A. J., Shannon, P. and Miller, D. (2008). 'Small Scale Wind Development in Aberdeenshire'. Report to Aberdeenshire Council and Scottish Natural Heritage. The Macaulay Institute and the University of Aberdeen. 68 pp.

Scott, A. J., Carter, C. E., Larkham, P., Reed, M., Morton, N., Waters, R., Adams, D., Collier, D., Crean, C., Curzon, R., Forster, R., Gibbs, P., Grayson, N., Hardman, M., Hearle, A., Jarvis, D., Kennet, M., Leach, K., Middleton, M., Schiessel, N., Stonyer, B. and Coles, R. (in press). 'Disintegrated Development at the Rural Urban Fringe: Re-connecting spatial planning theory and practice'. *Progress in Planning*.

Scottish Executive (2006). 'Sustainable Development: A Review of International Literature'. Scottish Executive Social Research, Edinburgh.

Scottish Government (2009). 'Final Results of the June 2009 Agricultural Census'. Scottish Government, Edinburgh. <http://www.scotland.gov.uk/Publications/2009/09/agriccensus 2009> (last accessed 8 January 2013).

Scottish Government (2011a). 'Scottish Government Rural Affairs and the Environment Portfolio: Strategic Research'. <http://www.scotland.gov.uk/Topics/Research/About/EBAR/ StrategicResearch/future-research-strategy/Themes/ThemesIntro> (last accessed 8 December 2012).

Scottish Government (2011b). 'Getting the best from our land: A Land Use Strategy for Scotland'. Scottish Government, Edinburgh.

Scottish Government (2011c) 'Getting the best from our land: A guide to Scotland's Land Use Strategy'. Scottish Government, Edinburgh.

Scottish Government (2011d). 'The Muirburn Code'. <http://www.scotland.gov.uk/Resource/ Doc/355582/0120117.pdf> (last accessed 5 December 2012).

Scottish Government (2012a). 'Scotland Rural Development Programme 2007 – 2013'. <http:// www.scotland.gov.uk/Topics/farmingrural/SRDP> (last accessed 8 December 2012).

Scottish Government (2012b). 'Scotland Performs'. <http://www.scotland.gov.uk/About/ Performance/scotPerforms> (last accessed 8 December 2012).

Scottish Mountain Bike Development Consortium (2009). 'The Sustainable Development of Mountain Biking in Scotland: A National Strategic Framework'. Forestry Commission Scotland.

Scottish Mountaineering Club (2012). 'Munro compleatists'. <http://www.smc.org.uk/Munros/ MunroistsCompleatists.php> (last accessed 8 January 2013).

Scottish Natural Heritage (2008). 'Public Perceptions of Wild Places and Landscapes in Scotland'. Market Research Partners, commissioned by SNH.

Selman, P. (ed.) (1988). *Countryside Planning in Practice: the Scottish experience*. Stirling University Press. 240 pp.

Shaw, H. and Whyte, I. (2008). 'Shifting ecosystem services through time in the North West uplands and implications for planning adaptation in the future'. *Aspects of Applied Biology* 85, pp. 99–106.

Simmons, I. G. (2003). *The moorlands of England and Wales. An environmental history 8000 bc–ad 2000*. Edinburgh University Press, Edinburgh. 414 pp.

Skerratt, S., Atterton, J., Hall, C., McCracken, D., Renwick, A., Revoredo-Giha, C., Steinerowski, A., Thomson, S., Woolvin, M., Farrington, J, and Heesen, F. (2012a). *Rural Scotland in Focus 2012*. Rural Policy Centre, Scottish Agricultural College, Edinburgh. 105 pp.

Skerratt, S., Farrington, J. and Heesen, F. (2012b) 'Next generation broadband in rural Scotland: mobilising, meeting and anticipating demand'. In: Skerratt, S., Atterton, J., Hall, C., McCracken, D., Renwick, A., Revoredo-Giha, C., Steinerowski, A., Thomson, S., Woolvin, M., Farrington, J, and Heesen, F., *Rural Scotland in Focus 2012*. Rural Policy Centre, Scottish Agricultural College, Edinburgh, pp. 70–85.

Smout, T. C. (2000). *Nature Contested: environmental history in Scotland and northern England since 1600*. Edinburgh University Press, Edinburgh. 210 pp.

SNH (2000). 'Policy Summary: National Scenic Areas'. Policy Note Series, Scottish Natural Heritage.

SNH (2012). 'Code of practice on deer management'. Scottish Natural Heritage, Battleby.

Sotherton, N., Tapper, S. and Smith, A. (2009). 'Hen harriers and red grouse: economic aspects of red grouse shooting and the implications for moorland conservation'. *Journal of Applied Ecology* 46 (5), pp. 955–60.

Staines, B. W., Balharry, R. and Welch, D. (1995). 'The impact of red deer and their management on the natural heritage in the uplands'. In: Thompson, D. B. A., Hester, A. J., and Usher, M. B. (eds), *Heaths and Moorland: Cultural landscapes*. HMSO, Edinburgh, pp. 294–308.

Stead, D. (2000). 'Unsustainable Settlements'. In: Barton, H. (ed.), *Sustainable Communities: The Potential for Eco-Neighbourhoods*. Earthscan, London, pp. 29–48.

Stockdale, A. and Barker, A. (2009). 'Sustainability and the multifunctional landscape: An assessment of approaches to planning and management in the Cairngorms National Park'. *Land Use Policy* 26 (2), pp. 479–92.

Stockdale, A. (2010). 'The diverse geographies of rural gentrification in Scotland'. *Journal of Rural Studies* 26, pp. 31–40.

Strachan, F. and Beck, C. (2008). 'Woodfuel, rural development and the natural heritage of the Highlands'. In: Galbraith, C. A. and Baxter, J. M. (eds), *Energy and the Natural Heritage*. TSO Scotland, Edinburgh, pp. 251–8.

Suckall, N., Fraser, E. and Quinn, C. (2009). 'How class shapes perceptions of nature: implications for managing visitor perceptions in upland UK'. In: Bonn, A., Allott, T., Hubacek, K. and Stewart, J. (eds), *Drivers of Environmental Change in Uplands*. Routledge, London and New York, pp. 393–403.

Sustainable Development Commission (2005). 'Wood fuel for warmth: a report on the issues surrounding the use of wood fuel for heat in Scotland'. Sustainable Development Commission, London.

Swales, V. (2009). 'The Lie of the Land: Future Challenges for Rural Land Use Policy in Scotland and Possible Responses'. A report commissioned by the RELU programme, January 2009.

Thirgood, S. J. and Redpath, S. M. (2008). 'Hen harriers and red grouse: science, politics and human–wildlife conflict'. *Journal of Applied Ecology* 45, pp. 1488–92.

Thompson, D. B. A., Nagy, L., Johnson, S. M. and Robertson, P. (2005). 'The nature of mountains: an introduction to science, policy and management issues'. In: Thompson, D. B. A., Price, M. F. and Galbraith, C. A. (eds), *Mountains of Northern Europe: Conservation, Management, People and Nature*. TSO Scotland, Edinburgh, pp. 43–56.

Thomson, S. (2010). 'How is Scotland's rural population changing?' In: Skerratt, S., Hall, C., Lamprinopoulou, C., McCracken, D., Midgley, A., Price, M., Renwick, A., Revoredo, C., Thomson, S., Williams, F. and Wreford, A., *Rural Scotland in Focus 2010*. Rural Policy Centre, Scottish Agricultural College, Edinburgh, pp. 9–17.

Tomlinson, R. W. (2010). 'Changes in the extent of peat extraction in Northern Ireland 1990–2008 and associated changes in carbon loss'. *Applied Geography* 30, pp. 294–301.

Tourism Intelligence Scotland (2010). 'Mountain Biking Tourism in Scotland – 7 Opportunities for Growth'.

UK National Ecosystem Assessment (2011). 'The UK National Ecosystem Assessment: Synthesis of the Key Findings'. UNEP–WCMC, Cambridge.

VisitScotland (2011). 'Scotland: The key facts on tourism in 2011'. <http://www.visitscotland.org/pdf/VS%20Insights%20Key%20Facts%202012_FINAL.pdf> (last accessed 8 January 2013).

van der Wal, R., Bonn, A., Monteith, D., Reed, M. S., Blackstock, K., Hanley, N., Thompson, D., Evans, M., Alonso, I. with Allot, T., Armitage, H., Beharry-Borg, N., Glass, J., McMorrow, J., Ross, L., Pakeman, R., Perry, S. and Tinch, D. (2011). 'Mountains, Moorlands and Heathlands'. In: UK National Ecosystem Assessment, UNEP-WCMC, Cambridge, pp. 105–60.

Ward, N. (2006). 'Rural Development and the Economies of Rural Areas'. In: Midgley, J. (ed.), *A New Rural Agenda*. Institute for Public Policy Research, London, pp. 46–67.

Warren, C. R. (2009). *Managing Scotland's Environment*. Second Edition. Edinburgh University Press: Edinburgh. 432 pp.

Wells, A. (2004). 'Sustainable Development on the Glenlivet Estate'. Glenlivet Estate Ranger Service, Tomintoul.

White, R. M., Fischer, A., Marshall, K., Travis, J. M. J., Webb, T. J., di Falco, S., Redpath, S. M. and van der Wal, R. (2009). 'Developing an integrated conceptual framework to understand biodiversity conflicts'. *Land Use Policy* 26, pp. 242–53.

Wightman, A. (1996). *Scotland's mountains: an agenda for sustainable development*. Scottish Wildlife and Countryside Link, Perth.

Wightman, A. (2004). 'Hunting and Hegemony in the Highlands of Scotland; A Study in the Ideology of Landscapes and Landownership'. Noragric Working Paper, No. 36. <http://www.andywightman.com/docs/noragric-wp-36.pdf> (last accessed 8 January 2013).

Williams, E. (2009) 'Preliminary exploration of the use of ecosystem services values in a regulatory context'. Environmental and Resource Economics Project Report for the Scottish Environment Protection Agency (SEPA).

Woods, M. (2005). *Rural Geography: Processes, Responses and Experiences in Rural Restructuring.* SAGE Publications, London. 352 pp.

Worrall, F., Reed, M. S., Warburton, J. and Burt, T. (2003). 'Carbon budget for a British upland peat catchment'. *Science of the Total Environment* 312 (1–3), pp. 133–46.

Yallop, A. R. and Clutterbuck, B. (2009). 'Land management as a factor controlling dissolved organic carbon release from upland peat soils 1: Spatial variation in DOC productivity'. *Science of the Total Environment* 407 (12), pp. 3803–13.

Zhang, N., Ricketts, T. H., Kremen, C., Carney, K. and Swinton, S. M. (2007). 'Ecosystem services and dis-services to agriculture'. *Ecological Economics* 64, pp. 253–60.

PART TWO

Perspectives from private landownership

CHAPTER THREE

The Scottish private estate

Annie McKee, Charles Warren, Jayne Glass and Pippa Wagstaff

INTRODUCTION

Landownership, land use and property rights have been extensively debated across Britain but, in Scotland, these debates are uniquely charged, both politically and socially. In the 1970s, the land reformer John McEwen argued passionately that 'the stranglehold of powerful, selfish, anti-social landlords [must be] completely smashed' (1977, 13) while Jim Hunter, another advocate of land reform, is on record as promising that 'when the sporting estate is dead and buried, I'll lead the dancing on its grave' (in Maxwell, 1998). One of the most prominent campaigners for reform, Andy Wightman (2010, 2, 4), has described Scotland as 'a country where, as a result of the theft of Scotland's commons . . . a tiny few hold sway over vast swathes of country'. He regards the so-called land question as 'very much unfinished business' and argues strongly for fundamental change. Such passionate views, and the situation that elicits them, are a product of the very particular history and nature of landownership in Scotland, a context which is strikingly different from elsewhere in Britain. This chapter outlines that history, focusing particularly on the evolution and character of private landownership, and then brings the story up to date by presenting the results of our recent research into the characteristics, motivations and viewpoints of those with a stake in land management today. Drawing on interviews with a range of expert commentators and a large survey of private landowners, both carried out in 2008 for the Sustainable Estates project, it throws contemporary light on some of the long-standing debates about the nature, place and value of private landownership in Scotland.

THE RISE OF THE SPORTING ESTATE

The current pattern of landownership in upland Scotland began to emerge in the mid eighteenth century with the establishment of individual property rights over former clan territory as a direct result of the Jacobite uprising of 1745–6 and its aftermath (Callander 1987; Devine 1995; Dunnett 2007). The traditional role of the clan chief as custodian was overtaken by economic considerations, and, as Prebble (1963, 16) describes, by the 'manners, demands and vanities of the south'. A Britain-wide shift under the control of a new class of landed gentry in the eighteenth and nineteenth centuries resulted in many Highland landlords clearing people from the land, often forcefully, in order to capitalise on the more profitable nature of sheep and cattle grazing which emerged as a result of agricultural improvements (Mackenzie 1998; Ali and Paradis 2006; Sellar 2006). The Highland Clearances, which involved the displacement and eviction of a large proportion of the rural population of the Highlands and Islands (particularly in the century from 1760), continue to generate debate among historians. There is no doubt, however, that this infamous period remains 'elemental in community awareness' (MacDonald 1998, 239), especially in many Highland areas, retaining powerful historic symbolism and often featuring prominently in contemporary political debate.

Furthermore, as a demonstration of the exploitative power exercised by some private landowners at that time, this era in Scottish history continues to generate negative sentiments towards private landownership and helps to explain the prominence of land issues in contemporary Scottish political debates (Devine 1995; Hunter 2006). It should be noted, however, that the rate, number and methods of eviction and displacement varied greatly across the Highlands: while many landowners demonstrated little or no social concern for those who lived and worked on their land, some re-accommodated 'cleared' communities (Richards 2000). As Devine (1995) explained, these differences derived from the variety of social and economic pressures affecting the region during the period, notably the scarcity of resources in Highland Scotland (compared to the coal reserves of the Lowlands, for example) and the suitability of the area for large-scale commercial pastoralism.

In 1852, Queen Victoria and her husband Albert launched a fashion for the Highlands when the couple acquired Balmoral Castle and Estate as a Highland base (Hunter 2006). This led to an increase in tourism and in the purchase of estates for sporting and other leisure purposes, particularly by the newly rich industrial magnates of the Victorian era. Sporting estates proliferated across the Highlands, and deer stalking, grouse shooting and fishing became hugely popular among the British upper classes (Warren 2009). The collapse of sheep prices in the 1870s released cheap land for sporting use which led to 60 per cent of land becoming shooting estates (Orr 1982). By 1873 half of Scotland's land was owned by just 118 people, and 50 per cent of the Highlands was in the hands of a mere fifteen landowners – a peak of concentrated ownership that prevailed for several decades (Armstrong and Mather 1983). Major landowners were in virtual control of Scotland's economy in the eighteenth and nineteenth centuries, and the relationship between landowner and tenant has been criticised for its exploitative nature (Bilsborough 1995; Cameron 2001):

> The hiring of labour by private landowners would give rise to injustices, whilst the landowner, simply through the fact of owning land often through inheritance, would become rich without adding one iota to the wealth of the community. (Bilsborough 1995, 4)

Though such landed power has led to infamous abuses in the past, it is also credited with progressive developments in Scotland's history: for example, the Lowland agricultural revolution in the eighteenth century (Smout 1997; Cameron 2001). During the late eighteenth and early nineteenth centuries, in conjunction with the rapid growth of Scotland's urban centres with industrialisation, Scotland's agricultural economy became highly commercialised and technically sophisticated, beginning with the removal of common land and the introduction of enclosure (Sutherland 1968). These developments are attributed to the capital and entrepreneurship injected by the Scottish landowning class (Caird 1983; Cameron 2001), and rural populations in many lowland regions consequently grew with the development of planned villages and industrial opportunities (Bird 1982). Known as the 'Improvers', and an important factor in the Age of Enlightenment, powerful private landowners were driven by notions of progress, efficiency and improvement (Turnock 1979; Sellar 2006). The notion of 'improvement', however, tended to highlight the utilisation of the estates' natural resources to the financial benefit of their owners while overlooking the ensuing dramatic declines in estate employment. In addition, the removal by some owners of common grazings undermined the rights of local communities, forcing them to rely completely on the landowning classes for their livelihoods (Sutherland 1968).

A salient argument often made in favour of private landownership is that the long-term perspective associated with ancestral ownership and the prospect of passing on land to future generations offer the potential to deliver long-term stewardship. Estate expenditure also tends to be focused in the local area, sustaining local businesses and communities. The historically paternalistic character of private landownership meant that lairds had a traditionally patriarchal relationship with tenants, with clear control over the availability of housing, employment and development opportunities on the estate. This generated hierarchical power relations: 'at the top was the Laird, who made it his business to know everyone else's business . . . and he was everyone's employer and landlord' (Bird 1982, 47). In the twentieth century, however, powers of local taxation and local justice administration were removed from local landowners via the centralisation of the planning and social welfare system after World War II (Bryden and Hart 2000). Nonetheless, landowners continue to play an important role in the development – or, in some cases, lack of development – of local businesses and communities in the uplands, influencing the size and distribution of an area's population and labour skills, as well as access to employment and land for development purposes. The contemporary roles and associated power of the private landowner are discussed in Chapter 5.

THE CURRENT DOMINANCE OF PRIVATE LANDOWNERSHIP

Scotland has the most concentrated pattern of private landownership in Europe, and possibly in the world (Lorimer 2000; Cahill 2001; Wightman 2001). Research by Wightman indicates that around 30 per cent of private rural land in Scotland is owned by only 115 owners, and just seventeen owners hold 10 per cent of the country (Wightman 2010; see also the Britain-wide figures produced by Cahill 2001). There are some 340 sporting estates in the Highlands and Islands which cover 2.1 million hectares and account for 43 per cent of all of the privately owned rural land (Higgins et al. 2002). This represents the largest concentration of land dedicated to game sport in Western Europe (Higgins et al. 2002). It is striking that, despite contrary economic forces, the high priority given to sporting land use and the recreational desires of the landowner have persisted since Victorian times (Strutt and Parker 2007). In general, a sporting estate is of necessity large, with hectares often counted in the thousands rather than the hundreds. The capital value of estates is based on the current value of available game, rather than on land area. One consequence of the high social value attached to game sport is that it can result in 'wildly inflated paper values for land which is amongst the poorest in Europe' (McCarthy 1998, 101).

Several researchers have contributed significantly to clarifying private landowner-ship patterns in Scotland. The most comprehensive landownership surveys in recent decades have been carried out by Millman (1969; 1970), McEwen (1977), Callander (1987) and Wightman (1996; 2012a). Several of these surveys were explicitly driven by an overt land-reform agenda, as the quotations at the beginning of the chapter exem-plify. Work by Clark (1981) identified a weakening of the oligopoly of landownership in many regions of Scotland between 1872/3 and 1970, illustrated by an overall reduc-tion in estate size and an increasing number of owners, though some estates expanded (Sutherland 1968; Clark 1981; Price et al. 2002). Wightman updated McEwen's seminal book *Who Owns Scotland?* in a volume of the same title (but without the question mark), and as a continuing project online.[1] Despite these efforts, it remains difficult to gather a complete picture of the extent of private landownership. Data on exactly 'who owns what and where' are inadequate, and researchers have had to rely on surveys from the early 1980s because of the lack of up-to-date, comprehensive data (Home 2009). In contrast to many other countries, Britain has no restrictions on who can buy land or the quantity of land purchased. Landownership information is held by the Inland Revenue, estate agents and landowners' representative groups. Official land registration in Scotland – the Land Registry and Register of Sasines – remains incomplete, voluntary to those properties not transacted since 1925, and expensive to access. All of these points make it a challenge to construct an accurate national picture of private landownership. The task would appear increasingly difficult with the growing variety of ownership and tenure mechanisms – for example, offshore financial investment trusts – and the significant number of foreign owners (Callander 1986; Mather 1995; Warren 2009; Home 2009).

ABSENTEE AND FOREIGN OWNERSHIP

Concurrent with the Victorian era, growth in the number of sporting estates owned by the upper class from the south of England, or 'new money' industrialists (Hunter 2006), there was a rise in absentee landownership in Scottish upland regions (Armstrong 1986; Warren 2009). Scottish estates have increasingly been bought and sold by owners who are not resident on the estate. By 1986, it was reported that four-fifths of owners derived their income from elsewhere and therefore needed to be elsewhere for much of the time.

Absenteeism, where the estate is not the owner's primary residence, is predominantly for recreational and/or investment purposes (Higgins et al. 2002; MacMillan et al. 2010). Most absentees have not been resident landowners previously and, in some cases, their period of ownership may last only a few years, either because they lose interest or because the substantial costs of running an upland estate become finan-cially unsustainable (Armstrong and Mather 1983). Non-economic motivations can outweigh economic reasons for land purchase, in particular for recreation (Petrzelka et al. 2013), and the rise of absenteeism has been mirrored by a growth in the number of estate management agencies – which are themselves generally 'absent' from the land they are contracted to manage (Armstrong 1986).

Absentee landownership has proved contentious over the past two centuries and remains difficult to measure in extent and impact. Whether a landowner is resident on his or her land throughout the year, or visits infrequently as an absentee, have an influence on the local community and development of the estate (Higgins et al. 2002; Petrzelka et al. 2013). Given the marginal economics of many landowning businesses, owners typically subsidise their estates from earnings made elsewhere, and this inflow of capital represents inward investment in remote rural areas (Price et al. 2002; Ali and Paradis 2006). Privately owned estates are often criticised for being 'unsustain-able' because of their dependence on substantial subventions by their owners but it is worth noting that 'inward investment' (whether public, corporate or private) is typi-cally welcomed as a positive contribution in other contexts whereas, in the context of upland Scotland, it often becomes tarred with the brush of absenteeism, transforming a positive into a negative (Warren 2009). On the other hand, there have been several high-profile cases in which the negative actions or neglect of absentee landowners have resulted in damage or degradation to the land resource, local community or tenantry (MacGregor 1988; McIntosh 2001; Bryden and Geisler 2007). The mismanagement of the Isle of Eigg prior to its purchase by the resident community is a well-known example (as described by McIntosh 2001). Absentee and/or foreign landowners have been widely criticised for leaving ownership and responsibility 'in the hands of people with no local roots and sense of belonging' (Dunion, in Paterson 2002, 72). Absentee landowners may be more resistant to change and innovative development than resident landowners who, in turn, may typically utilise the land more intensively than absentees (MacMillan et al. 2010). Overall, though, while many absentee landowners prioritise the sporting facilities of their estate over other estate functions, 'there is no simple correlation linking absentee owners with poor-quality management' (Warren 2009, 57) or, indeed, linking resident owners with enlightened management. No cat-egory of ownership is inherently or inevitably good or bad; exemplary or exploitative management can be practised by any kind of owner.

Debates surrounding absenteeism are closely linked with the discourse surround-ing foreign ownership. Foreign ownership quadrupled between 1970 and 1996, to around 6 per cent of private ownership (Wightman 1996). Information gathered for estates sold in 2004 showed that some 81 per cent of buyers were from the United Kingdom, 12 per cent from mainland Europe and the remaining 7 per cent from else-where (Strutt and Parker 2005). Interestingly, this differs from the situation in many other countries where the ownership of land by non-residents is severely restricted, and legal arrangements regulate the ownership and use of land in the public interest. In particular, Switzerland, Norway and Denmark have legislated regarding owner-occupation and the nationalities of landowners (Cahill 2001). Norwegian land law in the mid 1980s completely excluded anyone from owning land who was not resident on their land to ensure productive land use and a 'healthy rural climate' (McHattie 1986). With regard to the Scottish system, Cramb identifies as the 'single great flaw' the legal situation which 'allows anyone from anywhere in the world to buy huge areas of land and then fails to decree what can and cannot happen on the land' (2000, 21). Several of the most high-profile examples of irresponsible estate management have been under foreign ownership, and these have sparked calls for regulation of overseas landowner-ship (Wightman 1999). On the other hand, many foreign landowners manage their land responsibly and with a sense of stewardship (Price et al. 2002; Osborne 2007), the tenure of Paul van Vlissingen on Letterewe Estate being a salient example (Milner et al. 2002). Indeed, Smith (1983) made a strong argument for limited constraints on owner nationality in Britain because of the short supply of investment capital during the early 1980s – a situation that remains across Highland estates today and is implicitly supported by the lack of political will to restrict foreign landowners (Cramb 2000; Warren 2009). Constraints on foreign ownership could be achieved through capital taxation but the risk is that this would 'once again undermine the viability of let estates and re-start the trend towards greater owner occupation' (Munton 2009, S60). As illustrated, whether or not absenteeism is necessarily negative and whether Highland estates can retain financial viability without external investment are matters of ongoing discussion locally and nationally.

Private landownership in parts of the Highlands has sometimes been critically labelled as 'colonial' because the financial and management decision-making is undertaken outwith the local area (Mather 1999). Similar criticisms have also been levelled at NGOs and public agencies for being 'absentee' from the land that they own and manage (as discussed in Chapter 7). In the late twentieth century, several commentators challenged what they saw as the outdated and unjust characteristics of Scotland's landownership system. By raising public and political awareness of the dominance of private landownership in Scotland, they successfully put land reform on the political agenda, and this rapidly bore fruit in a newly devolved Scotland.

Twenty-first-century land reform

Scotland's land-reform legislation of 2003 continued a widespread and long-standing trend throughout Europe of property rights becoming progressively more restricted and qualified (Rodgers 2009; Pillai 2012). In addition to an economic case for Scottish land reform (MacMillan 2000), two key criticisms of the Scottish landownership situation were frequently articulated in support of calls for land reform. Firstly, con-cerns were raised regarding the unregulated nature of Scottish land sales, as outlined in the previous section. Secondly, the power of landowners was widely regarded as a hindrance to rural development by restricting access to resources and fostering a sense of 'powerlessness' among communities (Bryden 1997). This was stated as being compounded by the abuse of power by a small, but high-profile, minority who disregarded the aspirations of local people, fostering an enduring legacy of 'anti-landlordism' (Richards 2000; Cameron 2001). The process of 'social remembering' of the impact of the Highland Clearances remains a central aspect of community identity in the Highlands (Mackenzie 1998), leading to a loss of confidence by the public in private landowners (Munton 2009). As stated by Dingwall-Fordyce in his capacity as convener of the former Scottish Landowners Federation: 'the passionate memory of the Highland Clearances is likely to be a stumbling block to any effective advocacy of the landowner's case' (in Cameron 2001, 87).

The Land Reform (Scotland) Act (henceforth referred to as 'LRSA' or 'the Act'), passed in 2003, was an early, high-profile and deeply symbolic political objective of the recreated Scottish Parliament. The Labour government elected in 1997 had insti-gated the land-reform process by creating the Land Reform Policy Group (LRPG) with a remit to study the system of landownership and land management in Scotland. This group concluded that:

> Land reform is needed on grounds of fairness, and to secure the public good . . . The evidence indicates that present systems of land ownership and management in rural Scotland still serve to inhibit opportunities for local enterprise . . . The way landholdings are owned and managed can have a critical impact upon the land's ability to sustain rural populations. (LRPG 1998a, 2)

It recognised that diversity in landownership types and land use was the key to encouraging the greatest utilisation of rural development opportunities, as well as greater community involvement in land management (LRPG 1998b). This indicated a substantial reduction in the power of the landowner – a symbolic and significant step that aimed to increase the level of accountability of landowners, and poten-tially increase community ownership and management of land (Bryden and Hart 2000; Cameron 2001). The LRPG identified the main objective of land reform as the removal of 'land-based barriers to the sustainable development of rural communities' (LRPG 1999, 4), and it placed the concept of 'community' at the heart of its proposals (Bryden and Geisler 2007). The land-reform process was seen as 'a crucial opportu-nity to define clearly in contemporary terms the rights and responsibilities of private ownership in relationship to the public interest' (Callander 1998, 67).

The Act came into force in November 2004, concurrently with the Abolition of Feudal Tenure Etc. (Scotland) Act 2000. Unlike the rest of Europe and, indeed, England where feudalism was abolished in the twelfth century, Scotland retained ele-ments of this medieval land tenure system until the twenty-first century, and reform was widely perceived to be long overdue (Sellar 2006). The low level of public con-troversy surrounding the passage of this Act does not imply that it was insignificant in the landscape of private landownership in Scotland (Sellar, 2006). The abolition of feudalism altered the legal relationship between landowner and tenant (formerly 'superior' and 'vassal'), eliminating feudal burdens on tenants and removing a system that allowed individuals and institutions as 'superiors' to intervene in land ostensibly owned by others (Hunter 1995).

TABLE 3.1 Components of the Land Reform (Scotland) Act 2003
(Source: Warren and McKee 2011)

Components of the Land Reform (Scotland) Act 2003	Summary
Part 1 New statutory rights of public access to land and inland water	Provides statutory, non-motorised rights to be on land for: crossing an area; recreational purposes; relevant educational purposes; commercial purposes where these could be carried out under the right of access in a non-commercial way. Rights must be exercised responsibly and responsibilities are set out in the Scottish Outdoor Access Code (SOAC).
Part 2 Right to buy for local communities (a right of pre-emption)	A legally constituted community body (defined by postcode units and residency) has the right of first refusal to purchase previously registered local land when the owner decides to sell it, subject to certain criteria.[a] Chief among these criteria are the requirements that the purchase is in the public interest and that future use will be compatible with sustainable development.
Part 3 Absolute right to buy for crofting communities	This part of the Act is more radical as the initiative rests with the purchaser: if a crofting community chooses to exercise its right to buy croft land, the landowner is forced to sell it.

[a] Part 2 of the Act is stronger than this summary might imply in that it prohibits transfers of land to other parties, regulates the valuation of the land and provides a set period of time – six months – to complete the purchase, with recourse to the Lands Tribunal if necessary. It therefore places the community body in a strong bargaining position.

The LRSA is in three parts (see Table 3.1). Part 1 grants rights of responsible access to the Scottish countryside; Part 2 grants communities a right of pre-emption in the purchase of land entering the market; and Part 3 grants crofting communities an absolute right to buy croft land. Many commentators have described and evaluated the different sections,[2] and a lengthy explanation is not required here but the implications of the legislation, particularly of Parts Two and Three, emerge as key themes throughout the remaining chapters of this volume. Though the Act has significantly altered the landownership context, especially in crofting areas, it has been criticised in some quarters as being 'not radical enough' to reverse centuries of 'unfair' land distribution (Hunter 2012). Such critics regard it as merely a 'tokenistic nod', given that the inequalities of the current landownership system are considered by some to be central to wider rural social and economic injustices (Cameron 2001). In Paterson's view, 'it is not easy to find economic or social justifications for very large private landholdings, and . . . the high-priced but untaxed wild land known as sporting estates' (2002, 146). So how much will the Act actually change things?

IMPACTS OF LAND REFORM ON PRIVATE LANDOWNERSHIP

The longer-term effects of the Land Reform (Scotland) Act on private landownership and land management in upland Scotland are, of course, hard to predict. In the years

TABLE 3.2 Interviewees included in the survey[b] (2008)
(Adapted from: Warren and McKee 2011)

Description (sphere of interest and expertise)

- Landowning representative body; land management
- Landowner; academic; sustainable development
- Academic; historian
- Rural community development
- Government; community development
- Landowning representative body; land management
- Academic; sustainable development
- Government; land management
- Academic; community development

---

[b] Interviewees remain anonymous. Indicative descriptions are provided as a guide to illustrate the range of perspectives rather than specific affiliations.

that have passed since 2003, however, there has been much comment and speculation. Moreover, post-legislative scrutiny of the Act was commissioned by the Scottish Government in 2010 (Macleod et al. 2010) and again in 2012, indicating that there is a political appetite for further reform. In addition, an expert survey was carried out as part of the Sustainable Estates for the 21st Century project in 2008, in order to explore the impacts of the Act in more detail. The survey used interviews[3] with a select group of expert commentators who had extensive practical, academic and/or policy experience in the fields of land management, sustainable development and rural community development in Scotland (see Table 3.2). This section draws on the results of this survey as well as on a range of other published commentaries and research.

Since the LRSA was developed and passed, the 'land question' has become much more community oriented, with many private landowners expressing anxiety regarding increasing community powers (Higgins et al. 2002). Several successful community buyouts have taken place, stimulating local economic development and greater community cohesion (see Chapter 6). Despite the Act's high profile and the activity that it has generated, the new legal rights themselves have been exercised rather little. Part 2 of the Act makes provision for communities to purchase land, yet there has been very little change in the number of community-owned estates since the legislation was passed in 2003. By May 2012, of the ninety-five applications which had been approved, thirty-three had had the opportunity to proceed to purchase, and eleven had done so (Scottish Government 2012). Most of these buyouts consisted of small areas of land (under 2 hectares), with communities focusing on specific facilities or buildings rather than on substantial land areas. So though the total area purchased amounts to a sizeable 20,578 hectares, the greater part of that area is accounted for by a single large buyout, the estate purchased by the Assynt Foundation in 2005, with only three purchases exceeding 402 hectares (Scottish Government 2012).

Ironically, several of the best-known community purchases took place prior to the Act. The Assynt Crofters were the trailblazers of the recent community-ownership movement, their purchase of the North Lochinver Estate in 1992 demonstrating the potential for community ownership (MacPhail 2002). In the years that followed, high-profile examples of community purchases included the Isle of Eigg in 1997, Knoydart Estate in 1999, the Isle of Gigha in 2001, the North Harris Estate in 2003 and, in 2006, the 41,000-hectare South Uist Estate which is the largest community buyout to date (the last two are discussed in more detail in Chapter 6). Interestingly, the purchases of North Harris and South Uist estates did not use the Act's legal provisions but took place via agreements with the landowner, bypassing the complexities of the legal process (Warren and McKee 2011). MacLeod et al. (2010) identify the barriers which communities have to overcome when utilising the powers of the Act, including the administrative complexity, the potential to engender negative relations with local landowners, and the fact that community registration does not guarantee a result for the community because the owner is under no obligation to sell the registered land (Braunholz-Speight 2011). Arguably, these purchases might not have taken place if the Act was not on the statute book given that they were achieved without using the legal procedures provided by it; in this regard, so far the Act seems to be effecting change at least as much through leverage as through actual use (Slee et al. 2008).

During the final debate over the content of the Act, landowning interests predicted that the Act would 'depress land values, reduce investment and employment in rural Scotland, and put the natural heritage at risk by undermining the incentive for long-term, high-quality land management' (Warren and McKee 2011, 23). In our survey of expert commentators (see Table 3.2), one interviewee highlighted the fact that, because the powers of the Act have been little used since they came into force, many of the fears of landowners concerning the disruption of land markets and conventional land management remain unrealised. This commentator explains that this could be partly because the process is not easy, requiring commitment and patience, and also because the decisions regarding whether and when to sell, and what area(s) to put on the market remain with the owner. According to Home (2009, S105), the 'new tenure forms are making limited impact upon land-ownership patterns, but can be expected to grow'. This assertion was confirmed in the interviews, highlighting the small-scale impact of the Act on private landownership so far, especially outwith the crofting regions, thus assuaging the fears of the private landowning community. Though the low uptake was not unexpected, some interpret it as an implicit criticism of the law's effectiveness, and even of the need for legislation at all (Fletcher 2007). Concern has also been expressed that, in some cases, registration has been sought for negative reasons (for example, to block a development) rather than for the positive reasons envisaged in the Act, though such applications have typically been turned down by ministers (Pillai 2010).

Part 3 of the Act, the crofting right to buy, is its most radical aspect because it enables forced sales, and it is therefore unsurprising that landowning interests are deeply critical of this part, perceiving it as state-sponsored expropriation (Warren and McKee 2011). One interviewee directly countered this critique, arguing that the Act is the very opposite of 'belated revenge for the Clearances' or some kind of Marxist transfer of land into public ownership but is, instead, about 'enhancing the confidence of the Scottish people . . . with a view to boosting their prospects, giving them more control on their lives, and generally enabling them to go forward'. From this perspective, crofting buyouts may, indeed, deliver the promise of enhanced social justice and sustainability (Mackenzie 2006).

The true balance of benefits and costs of crofting-community ownership will only become apparent in the long term but, in the short term, it is already having a substantial impact in the crofting counties. In 2005, crofters on Lewis were the first to attempt to put the law into practice when they submitted applications to purchase the 25,000-hectare Galson Estate and the 10,000-hectare Pairc Estate, with others at Barvas and Soval following. By 2007, when the Galson Estate passed into community ownership, half the Western Isles of Scotland was owned by charities, communities and public bodies, with only two large privately owned estates remaining. Hamilton believes that the crofting community's right to buy could limit external investment if landowners 'live in constant fear of their estate being bought from under them' (Maxwell 2005). Indeed, an absence of crofters on a Highland estate may increase its value and make it more attractive to potential buyers, as a 'new owner will have total control over the estate and with no threat to their enjoyment' (Gibson, in Watson 2009).

It is difficult to separate the specific effects of land reform from other policy and market factors that affect socioeconomic well-being in Scotland (Slee et al. 2008). Returning to the interview survey, however, the legislation was regarded by one respondent as a landmark piece of legislation which, through both its powers and its existence, has the potential to have a dramatic impact on private landownership in upland Scotland. In contrast, another interviewee was disappointed at the 'watered down' nature of the Act, echoing concerns noted by land-reform activists, and another noted the failure of the Act to bring about any substantial transfer of ownership out of private hands. Wightman (2010) believes that the 2003 legislation fell short of what was needed, focusing his attention particularly on how the Act failed to address the large scale of existing landholdings, and how this could become a focus of land-reform demands in years to come. While recognising the significance of the Act, he sees it as just a beginning, and continues to call for a range of far-reaching reforms, including changes to land law and land policy, further promotion of community ownership, and the introduction of Land Value Taxation. In his latest book, published in 2010, Wightman raises the concern that private landownership has still not been subject to any critical analysis within government policy-making, a scenario that he deems worrying in the current economic climate where 'families are still homeless, the housed are burdened by record levels of debt, the young can't begin to imagine what it must be like to own their own home and young farmers have no prospect of getting hold of a farm' (p. 297).

The LRSA sought to alter the balance of power and, rather than addressing management practices, it focused on the 'ownership' of land. Ironically, perhaps, one of the impacts of the LRSA may be a reduction of investment in privately owned estates by their owners because of fears that the provisions of the Act will depress capital and/or land values. If inward investment is, indeed, reduced, then clearly there is the potential for negative knock-ons for local estate communities and for the stewardship of the environment. Again, however, opinions on this issue are divided. One interviewee was convinced that the Act has affected owners' investment decisions but another believed that the wider financial and economic situation is the primary consideration affecting such decisions and was sceptical that concern about land reform was a significant factor.

Furthermore, it may be questioned how the Act, and the ongoing land reform campaign, have influenced the engagement and involvement of private landowners and land managers in rural community initiatives, and, vice versa, the opportunities for rural communities to become involved in local land management. The Act could have a role in facilitating partnerships between landowners and communities but, at present, according to one interviewee, landowners are 'running around scared, and instead of engaging, they are hiding . . . [largely due to] fear of land reform'. Some landowners fear that greater engagement with their local communities may increase people's awareness of the provisions of the Act, thereby making them more inclined to utilise them. Arguably, however, this fear might be misplaced because effective engagement by the landowner is likely to defuse community desires to initiate land acquisition. On the other hand, the prospect of a referendum on Scottish independence in 2014 is generating renewed debate about land reform, reawakening fears among private landowners (Strutt and Parker 2012). Whether such fears are misplaced or not, another interviewee believes that the legislation has, in places, blocked attempts to create mutually beneficial partnerships between landowners and estate communities because fear and uncertainty impede engagement and collaboration between landowners and communities. Despite the many criticisms of the LRSA, Hunter (2012) is optimistic that the reinstatement of the Scottish Land Fund and the 2012 review of the Act may lead to a strengthening of the legislation and its practical value in supporting community landownership. The shift in the balance of power between owners and communities, which the Act has affected, and the impact of this on landowner–community dynamics has, until now, however, received little detailed attention, and this therefore is the focus of Chapter 5.

THE CONTEMPORARY PRIVATELY OWNED ESTATE

Today, estate-management activities continue generally to be dictated by a combination of the estate's natural resource base, the professional and personal motivations of the individual, group or organisation that owns and manages the estate, and the financial resources available (Kerr 2004). There have been two key studies that have profiled private landownership in twenty-first-century Scotland. First, Higgins et al. (2002) conducted a questionnaire survey in 2001 and 2002 of 172 sporting estates in the Highlands and Islands (also discussed by MacMillan et al. 2010). This dataset was combined with the results of some interviews carried out on a selection of estates. The study found that private landowners shared a set of core values, typically corresponding to traditional management aims and objectives (especially regarding sport), with 'little enthusiasm for change of any kind as it may undermine a comfortable and reassuring status quo' (MacMillan et al. 2010, 39). The survey data also presented the

Highland sporting estate as a 'relatively homogeneous phenomenon', characterised by:

> 15–20,000 acres with a hunting lodge; 8.5 full-time employees; owned by a man of significant but not immense wealth who lives elsewhere and owns land elsewhere; managed as a place to enjoy hunting and family holidays, [and] costing a 5-figure sum annually to balance the books. (Higgins et al. 2002: 5)

Second, a study carried out by Kerr in 2004 concluded that privately owned estates varied greatly according to the owner's objectives and estate resource opportunities. Though based on a small sample of ten estates (strongly criticised by Wightman 2004), this report found that 'small' and 'medium-sized' estates demonstrated shared objectives of improving economic and aesthetic estate value through diversified land-based business such as in-hand farming, property letting and commercial business activities, directly employing between five and ten people. Estates in these categories were also found to undertake a range of 'public good' and community activities. 'Large' estates displayed a similar profile but directly employed a much greater number of people and had a much larger property portfolio and greater commercial focus. Better defined as 'diversified land-based commercial businesses', large estates depended primarily upon income generated through property and commercial activities whereas small and medium-sized estates focused primarily on income generated by agricultural activities, followed by property letting. Significantly, this research found that estate expenditure, within all estate categories, was greatest within a 25-mile radius of the estate, demonstrating that privately owned estates play a role in supporting rural businesses and help to support 'the sustainability of their surrounding communities'.

A survey of private landowners was carried out as part of the Sustainable Estates project in 2008. The survey included questions on estate demographics, motivations for ownership, key values, and interpretations of the opportunities and barriers related to sustainable estate management. The survey was distributed to 245 members of the Scottish Rural Property and Business Association (a representative organisation for private landowners and rural businesses in Scotland, now called Scottish Land and Estates). Members who owned at least 2,000 hectares (4,942 acres) of land in Scotland were targeted. Eighty-four completed responses were received (a 34 per cent response rate) and these respondents owned a total of 726,911 hectares (1.8 million acres) of land in Scotland.[4] Table 3.6 shows the distribution of the sixty qualifying responses geographically by region. These estates cover a total area of 547,704 hectares (1,353,405 acres), with an average estate size of 9,128 hectares (22,557 acres). The majority of respondents (43 per cent) were aged between fifty-one and sixty-five; 18 per cent were under fifty years old. Eight of the landowners were female. All respondents reported having considerable involvement with the running of their estates. Table 3.4 shows the length of time that respondents have owned their estates (sometimes crossing generations).

TABLE 3.3 Distribution of survey responses by region

Estate location	Number of estates	Percentage respondents	Land area covered by total responses from region (ha)	Percentage of total area covered by respondents
Caithness and Sutherland	8	13.3	80,869.2	14.8
Ross and Cromarty	6	10.0	60,005.3	11.0
Highland/Inverness-shire	17	28.3	184,721.9	33.7
Moray	1	1.7	3,237.5	0.6
Aberdeenshire	3	5.0	53,418.5	9.8
Angus	1	1.7	3,035.1	0.6
Perthshire	9	15.0	38,160.5	7.0
Stirling/Tayside	2	3.3	15,917.1	2.9
East Lothian/East Central	2	3.3	22,210.9	4.1
Argyll	6	10.0	44,777.8	8.2
Strathclyde/Ayrshire	3	5.0	17,326.5	3.2
Dumfries and Galloway	1	1.7	2,023.4	0.4
Borders	1	1.7	22,000.0	4.0
Total	**60**	**100.0**	**547,703.7**	**100.0**

TABLE 3.4 Length of estate ownership (within the survey)

Length of time in family ownership	Number and percentage of estates
less than 25 years	14 (25)
26–50 years	9 (16)
51–100 years	15 (26)
101–200 years	9 (16)
less than 200 years	10 (17)
Cumulative figures for length of family ownership of estates	
less than 50 years	23 (40)
less than 100 years	38 (67)
less than or equal to 200 years	47 (83)
more than 200 years	10 (17)

Thirty-eight per cent of the respondents owned their estates personally while another 34 per cent stated that the estate was owned by their family. The remainder were owned predominantly by family trusts,[5] with one estate being held in a partnership and another in a limited company (though this last case appeared to be a purely legal arrangement to safeguard private ownership rather than as an investment by shareholders in a real corporate venture). Only 35 per cent of the landowners (twenty-one individuals) had purchased their estate. The majority of respondents (55 per cent) had inherited their estate and 8 per cent had acquired the land through other mechanisms: for example, as a gift or a combination of purchase and inheritance. More than 91 per cent of respondents, regardless of whether they had inherited or purchased their estates, wished to pass on their estate to their heirs.

Respondents who purchased their estate noted the importance of field sports, a place for a home, making a living or for their family, and environmental conservation in their motivations for acquiring their land. Those respondents who have inherited their estate consider environmental protection and sporting interests as a much lower priority in terms of motivations for retaining estate ownership, with over half the respondents noting 'stewardship' and responsibility as their key drivers in maintaining ownership, followed by expressions of privilege and enjoyment derived from landownership. Many respondents also described their justification for estate retention due to its status as a family asset and 'home', as well as for income generation and business pursuits. These themes of landowner motivation closely reflect those identified in 2002 by Higgins et al. (2002) and as described by MacMillan et al. (2010), suggesting that the passage of the LRSA has not (yet) influenced the motivations behind retention and purchase of upland estates.

A continuingly common characteristic of privately owned estates is the importance assigned to sporting land uses. Ninety-five per cent of respondents note deer stalking as an important estate enterprise, while 57 per cent and 45 per cent of respondents also indicate grouse shooting and pheasant shooting respectively. Though financial income can be generated from stalking, grouse shooting and fishing, almost two-thirds of privately owned estates involved in a study in 2002 were found to be unprofitable, with some sustaining annual losses running into six figures (Higgins et al. 2002). For example, the Letterewe Estate in Wester Ross was run at an annual net loss of over £130,000 at the turn of this century (Milner et al. 2002). For most owners, profit maximisation is a less important objective than keeping control of expenditure and protecting the capital value of the estate. As a result, cost may not always be a dominating factor in management decision-making. In our survey of private landowners, respondents were asked whether they expected all or parts of their estate to be profitable and whether they contributed funds from other sources either annually or occasionally for capital projects. Fifty-two per cent of respondents provide external funding from other sources to run the estate and 63 per cent contribute funds from outwith the estate towards capital projects on it. These figures indicate that estate economic viability is frequently reliant on outside funding. Regarding the significance of external funding (for both annual running costs and capital projects), respondents' comments fall into three categories. The first group highlight the need for regular external financial assistance. For a second group, the injection of external funds is not regular, and variable sums (often small) are borrowed or sourced when necessary. The third group of respondents explain that their aim is to become self-sufficient, and that the main external funding source is from grant aid.

A recent study of forest ownership in four areas of Scotland revealed that 46 per cent of private forest owners are absentees (accounting for 56 per cent of privately owned forest area) and that 69 per cent of these absentees do not live in Scotland (Wightman 2012b). In our survey, the overall average time spent on the estate by all respondents was seven months per annum, with 53 per cent indicating that their estate was their principal place of residence. The remaining 47 per cent classified themselves as absentee which corresponds with the level of absenteeism indicated by

Wightman (2012b). Several absentee landowners commented that they intended to retire to their estates in due course or when finances permitted.

In many places, private landholdings play a pivotal role in the rural economy, delivering a range of public benefits, such as conservation, public access, local employment and affordable rural housing, and also encouraging entrepreneurship in the local community (Mc Morran 2009). Box 3.1 shows the range of 'other' estate activities highlighted in our survey. In the sample, over 50 per cent of respondents also identified grouse shooting, tenanted farming, commercial forestry and tourism or holiday accommodation as income providers in the estate business. These may be considered 'traditional' estate enterprises, with arguably 'newer', innovative income sources, such as wind, biomass and other renewable energy production, being noted by fewer than 20 per cent of respondents. This demonstrates that, while many upland estates are diversifying, traditional estate enterprises remain central to income generation.

Box 3.1 Explanations of 'other' estate activities
(25 per cent of survey responses)

Publicly open gardens and plant nurseries	Forestry management and contracting
Wider game-bird shooting	Hotels
Fish farming	Caravan parks
Crofting	Golf courses
Conservation work	Public access, walking and cycling
Film, photo and wedding locations	Hospitality, catering and retail
Renewable energy	

Estate enterprise continues to provide job opportunities in upland Scotland, providing a livelihood for tenant farmers, crofters, foresters, gamekeepers, ghillies,[6] housekeepers, cooks and estate-maintenance employees (Ward 2006). Estate jobs have been found to be a significant source of employment in remote rural areas (Fraser et al. 2012; Mc Morran 2009). Some large estates [under 20,000 hectares (50,000 acres)], however, may only employ a 'handful' of people retained on low wages, probably as a result of estates being retained for 'pleasure' rather than 'business' (Cramb 2000), and there has been an increase in service- and tourism-based employment, for example on advantageously located estates such as Rothiemurchus and Atholl Estates. A tendency to integrate estate enterprises is visible through the process of replacing numerous specialist-sector staff with only a few general workers, capable of undertaking farming, forestry and sporting tasks. Our survey found that the average number of employees on each estate equated to 8.6 full-time equivalents (FTEs). This was based on a total of five hundred employees, ranging from none to 116, on sixty respondent estates, with an additional average of 2.5 FTE seasonal workers per estate, ranging from none to fifteen. This level of employment, though small in absolute terms, is often socioeconomically significant in the remote, rural communities of the Scottish uplands – a point discussed further in Chapter 5.

Private landowners have historically played a pivotal role in providing and allocating housing to the estate community. The example of Luss Estate in Argyll presented an insight into the importance of the private landowner in meeting local housing need, where the tradition was to let houses only to estate employees and their families, and never to evict widows and dependants (Bird 1982). This housing-provision model resulted in a stable and cohesive community, though also very insular and resistant to change (Bird 1982). Bowler and Lewis (1987) highlight two main landowner types: a minority seeking to build settlements in order to grow capital wealth, and a majority to whom the sale of development land is a necessity but who strive to maintain the 'existing social and material fabric' of local villages, with such landowners concerned to alleviate rural housing problems. Satsangi (2005) expands on and updates this typology with quantitative data from a large-scale questionnaire of private landowners involved in rural housing provision, revealing that the majority of landowner respondents strove to cover the costs of their rented housing, and to produce a profit if possible, with support for the local community a secondary consideration. Box 3.2 summarises survey findings related to rural housing in our research. Case studies highlight the common practice of private landowners letting at below market rent levels to 'local' people, and the aspiration of landowners to maintain 'affordable' rents because of an awareness of tenants' regular dependence on employment that is seasonal, part-time and low paid.

Box 3.2 Survey findings relating to private landowners and rural housing (based on responses from 60 estates)

- Residential property letting is a feature of 77 per cent of estate businesses.
- Affordable housing is considered vitally important by 12 per cent of respondents, important by 27 per cent, a little important by 22 per cent and not at all important by 13 per cent (with a further 23 per cent considering it 'not relevant').
- 22 per cent of respondents have sold land to developers for house building, and 33 per cent have sold land to individuals for housing.
- 35 per cent of respondents currently provide housing for estate employees, with 20 per cent hoping to do this in the future.
- 37 per cent of respondents provide housing to the local community at affordable rents, though the attempts of 3 per cent are failing. 13 per cent have plans for more rented affordable housing.

CONCLUSION

Despite the 2003 land-reform legislation, private landownership remains dominant in upland Scotland and is set to remain so for the foreseeable future, notwithstanding the increasingly pluralistic pattern of ownership types that has been emerging. Arguably, the privately owned estate has not changed significantly since the development of the current traditions in the eighteenth and nineteenth centuries, as outlined at the beginning of this chapter. As the findings of our survey illustrate, sporting activity and its associated landscape management remain at the core of most private estates despite significant diversification and the desire to generate economically viable rural businesses. The survey findings also illustrate extensive family ownership, often over several centuries, and the key motivations for ownership including a sense of responsibility and stewardship. Less flattering is the critical view of owners' motivations that estate ownership is as much for tax reasons as for game sport, a persistent critique which contributes to negative perceptions of private landownership and its legitimacy (Monbiot 2012). This is discussed further in the next chapter. Given the dominance of private landownership, concerns continue to be raised regarding access to resources and rural socioeconomic development, questioning the balance of power, management practices and ultimately, accountability.

Media reporting on 'who owns Scotland' generates a mixture of applause and outcry, and there has been a reawakening of popular concern regarding the role and status of private landownership. The Land Reform (Scotland) Act 2003 reflected the heightened prominence of landownership and management issues in post-devolution politics but, in much of rural Scotland, the Act may have a somewhat limited and slow-burning impact, inevitably frustrating those who are impatient for change and reassuring those who champion the benefits of private ownership. But, in the crofting areas, it has already had transformative effects in less than a decade by changing the balance of power between landowners and local communities. Further and more radical legislation may be recommended by the Land Reform Review Group (LRRG), established in 2012 with a remit to find ways of increasing the number of people who 'have a stake in the ownership, governance, management and use of land'.[7] A sense of uncertainty, mistrust and fear about future legislation was already present among interviewees in the Sustainable Estates project, and the fact that the First Minister believes that the LRRG should aim to deliver 'radical change' and make 'more innovative proposals' will do nothing to allay such feelings (Scottish Government News 2012). The existence of the Act and the prospect of further reform may in themselves be sufficient to initiate more dialogue between landowners and local communities, enabling them to work together to find mutually beneficial solutions (as explored in Chapter 5). In a sense, all types of landowners have been put on notice that the needs and aspirations of local people cannot be ignored.

There is, without doubt, a continuing trend towards greater public accountability of private landownership, and of land management in general. Arguably, the most important challenge is how best to structure ownership and associated management practices to facilitate the delivery of rural sustainability, safeguarding private and public interests. It is the balance of these two aspects that perhaps needs further scrutiny. What mixture of rights and responsibilities can facilitate the development of a healthy society and healthy environment? Which principles should inform any rebalancing of these rights, and who should decide on the principles and values in the first place? What type of landownership best delivers these principles 'on the ground'? There are no simple or agreed answers to these questions nor will any one solution be appropriate in all places. It is difficult even to identify simple correlations between types of estate ownership, land use and the enactment of rights and responsibilities, not least because there are numerous factors other than ownership that influence environmental and social outcomes. The motivations, desires and decisions of land-owners play an important part in influencing standards of environmental, economic and social responsibility, and this is the focus of the next chapter.

NOTES

1. http://www.whoownsscotland.org.uk/
2. For more detailed reviews of the Act, see Mackay 2007; Warren 2009; Macleod et al. 2010; Warren and McKee 2011.
3. Further detail regarding interview methodology may be found in Warren and McKee (2011).
4. Of the eighty-four responses, sixty fitted the criteria developed for the whole Sustainable Estates project: estates larger than 2,000 hectares, predominantly upland [over 185 metres (607 feet)] and with multiple land uses and estate activities.
5. Various structures are often put in place to provide asset protection to reduce the risks of expropriation through divorce, bankruptcy or the simple lack of business sense of the heir, as well as to mitigate Capital Gains and Inheritance Tax liabilities.
6. A 'ghillie' is the traditional term for the estate employee responsible for river bank manage-ment and recreational fishing on the estate, as well as usually also supporting other hunting activities.
7. http://www.scotland.gov.uk/About/Review/land-reform/ReviewGroup

REFERENCES

Ali, S. H. and Paradis, R. (2006). 'The Politics of Land Tenure in Scotland: A Comparative Study of Public, Private and Community Arrangements'. Working Paper.

Armstrong, A. M. (1986). 'Absentee Landowners in the Highlands'. *Scottish Forestry* 40 (2), pp. 84–6.

Armstrong, A. S. and Mather, A. S. (1983). *Land Ownership and Land Use in the Scottish Highlands*. Department of Geography, University of Aberdeen.

Bilsborough, S. (1995). 'A Hidden History: Communal Land Ownership in Britain.' *ECOS* 16, pp. 3–4.

Bird, S. (1982). 'The impact of estate ownership on social development in a Scottish rural com-munity'. *Sociologia Ruralis* 22, pp. 36–48.

Bowler, I. R. and Lewis, G. J. (1987). 'The decline of private rented housing in rural areas: a case study of estate villages in Northamptonshire'. In: Lockhart, D. and Ilbery, B. (eds), *The Future of the British Rural Landscape*. Geo Books, Norwich, pp. 115–36.

Braunholtz-Speight, T. (2011). 'Post-legislative Scrutiny of the Land Reform (Scotland) Act 2003'. Paper delivered at 'Whose Economy? Whose Environment?' UWS and Oxfam Seminar Series, Inverness College UHI, 25 March 2011.

Bryden, J. M. (1997). 'The land question in Scotland'. *Scottish Association of Geography Teachers Journal* 26, pp. 18–24.

Bryden, J. and Geisler, C. (2007). 'Community-based land reform: Lessons from Scotland'. *Land Use Policy* 24, pp. 24–34.

Bryden, J. and Hart, K. (2000). 'Land Reform, Planning and People: An Issue of Stewardship?' In: Holmes, G. and Crofts, R. (eds), *Scotland's Environment: The Future*. Tuckwell Press, East Linton, pp. 104–18.

Cahill, K. (2001). *Who Owns Britain: the hidden facts behind landownership in the UK and Ireland*. Canongate, Edinburgh. 464 pp.

Caird, J. B. (1983). 'Patterns of Rural Settlement, 1700–1850'. In: Clapperton, C. M. (ed.), *Scotland: A New Study*. David and Charles, Newton Abbott, pp. 128–53.

Callander, R. (1986). 'Background paper: The Law of the Land'. In: Hulbert, J. (ed.), *Land: Ownership and Use*. Andrew Fletcher Society, Longforgan, Dundee, pp. 1–11.

Callander, R. (1987). *A Pattern of Landownership in Scotland, with particular reference to Aberdeenshire*. First Edition. Haughend Publications, Finzean. 155 pp.

Callander, R. (1998). *How Scotland is Owned*. Canongate, Edinburgh. 226 pp.

Cameron, E. A. (2001). "Unfinished business': The Land Question and the Scottish Parliament'. *Contemporary British History* 15 (1), pp. 83–114.

Clark, G. (1981). 'Some secular changes to landownership in Scotland'. *Scottish Geographical Magazine* 97, pp. 27–36.

Cramb, A. (2000). *Who owns Scotland now? The use and abuse of private land*. Mainstream, Edinburgh. 206 pp.

Devine, T. M. (1995). *Exploring the Scottish Past: Themes in the History of Scottish Society*. Tuckwell Press, East Linton. 271 pp.

Dunnett, A. (2007). *The Canoe Boys*. Neil Wilson Publishing Ltd, Glasgow. 224 pp.

Fletcher, R. (2007). 'Buying by rights'. *Scottish Field* 104 (10), pp. 59–60.

Fraser, P., MacKenzie, A. and MacKenzie, D. (2012). 'The economic importance of red deer to Scotland's rural economy and the political threat now facing the country's iconic species'. Scottish Gamekeepers Association, March 2012.

Higgins, P., Wightman, A. and MacMillan, D. (2002). 'Sporting estates and recreational land-use in the Highlands and Islands of Scotland'. Report for Economic and Social Research Council.

Home, R. (2009). 'Land ownership in the United Kingdom: Trends, preferences and future challenges'. *Land Use Policy* 26 (1), S103–S108.

Hunter, J. (1995). 'Towards a Land Reform Agenda for a Scots Parliament'. The 2nd John McEwen Memorial Lecture. <http://www.caledonia.org.uk/land/hunter.htm> (last accessed 19 December 2012).

Hunter, J. (2006). 'Fonn 's dùthchas = Land and legacy'. NMS Enterprises Limited Publishing, Edinburgh.

Hunter, J. (2012). 'This land is our land'. *The Sunday Herald*, 11 March 2012. <http://www.heraldscotland.com/comment/columnists/this-land-is-our-land.16948642> (last accessed 8 May 2012).

Kerr, G. (2004). 'The contribution and socio-economic role of Scottish estates: summary report'. Scottish Agricultural College, Penicuik. 30 pp.

Lorimer, H. (2000). 'Guns, Game and the Grandee: The Cultural Politics of Deerstalking in the Scottish Highlands'. *Cultural Geographies* 7 (4), pp. 403–31.

LRPG (1998a). 'Identifying the Problems'. Land Reform Policy Group. Scottish Office, Edinburgh.

LRPG (1998b). 'Identifying the Solutions'. Land Reform Policy Group. Scottish Office, Edinburgh.

LRPG (1999). 'Recommendations for Action'. Land Reform Policy Group, Scottish Office, Edinburgh.

McCarthy, J. (1998). *An Inhabited Solitude: Scotland, Land and People*. Luath Press, Edinburgh. 155 pp.

MacDonald, F. (1998). 'Viewing Highland Scotland: ideology, representation and the "natural heritage"'. *Area* 30 (3), pp. 237–44.

MacGregor, B. D. (1988). 'Owner Motivation and Land Use on Landed Estates in the North-west Highlands of Scotland'. *Journal of Rural Studies* 4 (4), pp. 389–404.

McHattie, A. (1986). 'Crofting – is there a future?' In: Hulbert, J. (ed.), *Land: Ownership and Use*. Andrew Fletcher Society, Dundee, pp. 44–9.

McIntosh, A. (2001). *Soul and Soil: People versus Corporate Power*. Aurum Press, London. 384 pp.

Mackay, J. W. (2007). 'New legislation for outdoor access: a review of Part 1 of the Land Reform (Scotland) Act 2003'. *Scottish Affairs* 59, pp. 1–29.

MacKenzie, A. F. D. (1998). 'The Cheviot, The Stag . . . and the White, White Rock? Community, identity, and environmental threat on the Isle of Harris'. *Environmental Planning D: Society and Space* 16, pp. 509–32.

Mackenzie, A. F. D. (2006). 'A working land: crofting communities, place and the politics of the possible in post-Land Reform Scotland'. *Transactions of the Institute of British Geographers* 31 (3), pp. 383–98.

Macleod, C., Braunholtz-Speight, T., Macphail, I., Flyn, D., Allen, S. and Macleod, D. (2010). 'Post Legislative Scrutiny of the Land Reform (Scotland) Act 2003'. Final Report, September 2010.

MacMillan, D. C. (2000). 'An economic case for land reform'. *Land Use Policy* 17, pp. 49–57.

MacMillan, D. C., Leitch, K., Wightman, A. and Higgins, P. (2010). 'The management and role of Highland sporting estates in the early 21st Century: the owner's view of a unique but contested form of land use'. *Scottish Geographical Journal* 126 (1), pp. 24–40.

Mc Morran, R. (2009). 'The benefits and impacts of the grouse shooting industry from the rural community perspective: a case study of the Strathdon and Tomintoul communities in the Cairngorms National Park'. Centre for Mountain Studies, Perth, College UHI. 57 pp.

MacPhail, I. (2002). 'Relating to land: the Assynt Crofters Trust'. *ECOS* 23 (1), pp. 26–35.

Mather, A. S. (1995). 'Rural land occupancy in Scotland: resources for research'. *Scottish Geographical Magazine* 111 (2), pp. 127–31.

Mather, A. (1999). 'The Moral Economy and Political Ecology of Land Ownership'. Paper presented to the Land Reform in Scotland Conference, School of Planning and Housing, Edinburgh College of Art/Heriot-Watt University, Edinburgh, 19 October 1999.

Maxwell, F. (2005). 'Land reform hopes and fears unrealised', *The Scotsman*, 19 December 2005. <http://news.scotsman.com/topics.cfm?tid=694&id=2427952005> (last accessed 29 October 2012).

Millman, R. (1969). 'The Marches of the Highland Estates.' *Scottish Geographical Magazine* 85, pp. 172–81.

Millman, R. (1970). 'The Landed Estates of Northern Scotland'. *Scottish Geographical Magazine* 86, pp. 186–203.

Milner, J., Alexander, J. and Griffin, C. (2002). *A Highland Deer Herd and its Habitat*. Red Lion House, London. 367 pp.

Monbiot, G. (2012) 'Mythologists of the Glen'. *George Monbiot*, 2 March 2012. <http://www.monbiot.com/2012/03/02/mythologists-of-the-glen> (last accessed 2 June 2012).

Munton, R. (2009). 'Rural land ownership in the United Kingdom: Changing patterns and future possibilities for land use'. *Land Use Policy* 26S, S54–S61.

Orr, W. (1982). *Deer Forest, Landlords and Crofters: the Western Highlands in Victorian and Edwardian times*. John Donald, Edinburgh. 226 pp.

Osborne, J. M. (2007). 'The role of land management in the uplands'. Moors for the Future Partnership Conference, 2007.

Paterson, A. (2002). *Scotland's Landscape: Endangered Icon*. Polygon, Edinburgh. 256 pp.

Petrzelka, P., Ma, Z. and Malin, S. (2013). 'The elephant in the room: Absentee landowner issues in conservation and land management'. *Land Use Policy* 30, pp. 157–66.

Pillai, A. (2010). 'Sustainable Rural Communities? A legal perspective on the community right to buy'. *Land Use Policy* 27, pp. 898–905.

Pillai, A. (2012). 'Land Law'. In: Mulhern, M. (ed.), *Scottish Life and Society: A compendium of Scottish Ethnography*: Volume 13. Institutions of Scotland – The Law. University of Edinburgh, Edinburgh, pp. 367–88.

Prebble, J. (1963). *The Highland Clearances*. Penguin, Harmondsworth. 336 pp.

Price, M. F., Dixon, B. J., Warren, C. R. and Macpherson, A. R. (2002). *Scotland's Mountains: Key Issues for their Future Management.* Scottish Natural Heritage, Battleby. 90 pp.

Richards, E. (2000). *The Highland Clearances: People, Landlords and Rural Turmoil.* Birlinn, Edinburgh. 379 pp.

Rodgers, C. (2009). 'Property rights, land use and the rural environment: a case for reform'. *Land Use Policy* 26S, S134–S141.

Satsangi, M. (2005). 'Landowners and the structure of affordable housing provision in rural Scotland'. *Journal of Rural Studies* 21 (3), pp. 349–58.

Scottish Government (2012). 'Overview of Evidence on Land Reform in Scotland'. Agriculture, Fisheries and Rural Affairs. <http://www.scotland.gov.uk/Resource/0039/00397682.pdf> (last accessed 12 December 2012).

Scottish Government News (2012). 'Radical rethink on land reform underway', 24 July 2012. <http://www.scotland.gov.uk/News/Releases/2012/07/Land-Reform24072012> (last accessed 12 December 2012).

Sellar, D. W. H. (2006). 'The great land debate and the Land Reform (Scotland) Act 2003'. *Norsk Geografisk Tidsskrift/ Norwegian Journal of Geography* 60 (1), pp. 100 – 9.

Slee, B., Blackstock, K, Brown, K., Dilley, R., Cook, P., Grieve, J. and Moxey, A. (2008). *Monitoring and evaluating the effects of land reform on rural Scotland – a scoping study and impact assessment.* Scottish Government Social Research, Edinburgh. 210 pp.

Smith, E. A. (1983). 'Scotland's Future Development'. In: Clapperton, C. M. (ed.), *Scotland: A New Study.* David and Charles, Newton Abbot, pp. 281–99.

Smout, T. C. (1997). *Scottish Woodland History.* Scottish Cultural Press, Edinburgh. 215pp.

Strutt and Parker (2005). 'Scottish Estates Review'. Strutt and Parker, Edinburgh.

Strutt and Parker (2007). 'Scottish Estates Review'. Strutt and Parker, Edinburgh.

Strutt and Parker (2012). 'Scottish sporting estate market open for business'. *Strutt and Parker Rural News,* 7 February 2012. <http://www.struttandparker.com/news/rural-property/scottish-sporting-estate-market-open-for-business> (last accessed 8 May 2012).

Sutherland, D. (1968). *The Landowners.* Anthony Blond, London. 180 pp.

Turnock, D. (1979). 'Glenlivet: two centuries of rural planning in the Grampian uplands.' *Scottish Geographical Magazine* 95, pp. 165–81.

Ward, N. (2006). 'Rural Development and the Economies of Rural Areas'. In: Midgley, J. (ed.), *A New Rural Agenda.* Institute for Public Policy Research, London, pp. 46–67.

Warren, C. R. (2009). *Managing Scotland's Environment.* Second Edition. Edinburgh University Press, Edinburgh. 490 pp.

Warren, C. R. and McKee, A. (2011). 'The Scottish Revolution? Evaluating the impacts of post-devolution land reform'. *Scottish Geographical Journal* 127 (1), pp. 17–39.

Watson, J. (2006). 'Spirit of the Clearances lives on in estate's £2.5m price tag'. *The Scotsman,* 26 February 2006. <http://www.scotsman.com/news/scottish-news/top-stories/spirit-of-clearances-lives-on-in-estate-s-163-2-5m-price-tag-1-1409648> (last accessed 12 December 2012).

Wightman, A. (1996). *Who owns Scotland?* Canongate, Edinburgh. 237 pp.

Wightman, A. (1999). *Scotland: land and power – the agenda for land reform.* Luath Press, Edinburgh. 126 pp.

Wightman, A. (2001). 'Land Reform Draft Bill Part II – Community Right-to-Buy'. Caledonia Briefing No. 3. Caledonia Centre for Social Development.

Wightman A. (2004). 'The Contribution and Socio-Economic Role of Scottish Estates – A report prepared by the Scottish Agricultural College: A Critique by Andy Wightman'. Version 2.2, <http://www.andywightman.com/docs/critique_040115.pdf> (last accessed 12 December 2012).

Wightman, A. (2007). 'Land reform: an agenda for the 2007–2011 Scottish Parliament'. <http://www.andywightman.com/docs/landreform2007-final.pdf> (last accessed 12 December 2012).

Wightman, A. (2010) *The Poor Had No Lawyers: Who owns Scotland and how they got it*. Birlinn, Edinburgh. 320 pp.

Wightman, A. (2012a). 'Who Owns Scotland'. <http://www.whoownsscotland.org.uk/> (last accessed 12 December 2012).

Wightman, A. (2012b). 'Forest Ownership in Scotland: A Scoping Study'. Forest Policy Group, Perthshire.

CHAPTER FOUR

What motivates private landowners?

Pippa Wagstaff

INTRODUCTION

The previous chapter explained that a large proportion of the Scottish uplands is owned privately and that private landowners can exert a strong influence on land-use changes. The attitudes and aspirations of private landowners are therefore pivotal in implementing sustainable upland management in Scotland, and an understanding of the motivations of private landowners is essential for the design of effective land-use policy. This chapter explores the motives behind the actions of a sample of private landowners in upland Scotland. Building on the results presented in Chapter 3, data gathered on a sample of privately owned estates provide insights into the motivations of the selected landowners, together with a review of the extent to which their ambitions are aligned with public goals for sustainability.

LANDOWNERS' MOTIVATIONS AND ATTITUDES

Human behaviour tends to be goal oriented, with the goals being constructed from a combination of personal beliefs, values, attitudes and motives. The relative importance of each goal varies between individuals, and within each individual according to their time, circumstance, and social position (Gasson and Errington 1993). For example, younger landowners have been found to be more entrepreneurial because they are 'rich in time' while older landowners become more risk averse particularly where there is a prime objective of succession (Hutson 1987). When buying an estate, wealthy individuals are willing to pay large sums to gain access to the non-monetary benefits of landownership (such as leisure or as a 'hideaway', as discussed in Chapter 3), placing a value on land far in excess of its economic value (Denman 1965). Lifestyle buyers looking for an alternative to city living (Mather et al. 2006) and those seeking to enhance their social status through 'conspicuous leisure' (Veblen 1908) are continuing to embrace the Victorian fashion for country sports, reinforcing the link between power, money and game (Heley 2010); they are buying into the dream and the rich tapestry of tradition that goes with it. Such sporting interests are often criticised, however, for taking precedence over social and economic development (MacMillan et al. 2010). Logically, wealthy landowners who are not driven primarily by economic motivations are likely to place greater emphasis on their own expressive and social values than those, driven by economic need, who are dependent on earning a living from their land. For example, some research has found that financial goals are incompatible with sporting objectives (Armstrong 1980). Inheritance of an estate has also been shown to create values of attachment (Ilbery 1983; Moxnes Jervell 1999); involvement from an early age internalises family values (Armstrong 1980; Gasson and Errington 1993; Gray 1998; Riley 2009), often leading to a high level of continuity of decision-making.

There is no standard methodology for assessing the optimum balance of land uses, not least because what is judged 'optimum' will depend on aims and objectives, but the sustainability assessment tool presented in Chapter 8 represents one way forward. Though economic considerations often take precedence for private landowners (Miller et al. 2009), it is important to recognise that government policies based on financial incentives are not always successful (MacGregor 1988; Petrzelka et al. 2012) because not all landowner decisions are made solely on the basis of economic rationality (Wallace 1998; Schneider et al. 2010; Burton and Paragahawewa 2011); the pursuit of rural lifestyles, field sports or environmental goals are also frequently significant drivers (Armstrong 1980; MacGregor 1988; MacGregor and Stockdale, 1994; MacMillan et al. 2010). As a result, decision-making tends to be highly individual, strongly influenced by tradition, fashion, or the specific beliefs of the owner.

INVESTIGATING OWNERS' PRIORITIES ON SELECTED ESTATES

Eleven privately owned estates were studied in detail to develop a greater understanding of the motivations of their owners. Using the responses to the survey of private landowners (discussed in the previous chapter), estates were selected to reflect a cross-section of 'types' based on the degree of importance each respondent had assigned respectively to economic, social and environmental goals, and whether the landowner was 'resident' or 'absentee'. Owners were regarded as 'resident' if the estate was their principal place of residence.

Based on owners' responses, estates were categorised according to (i) the dominant motivations identified in landowner survey responses, (ii) whether resident or absentee, and (iii) the mode of acquisition (inherited or purchased), as shown in Figure 4.1. Certain combinations, according to these categories, proved hard to find; they may not, in fact, exist. Most of the estates studied had resident landowners who had inherited their estates. It was possible to work with only one absentee landowner who had purchased his estate, though Estate 11 also fitted within this category.[1] Such landowners may, by their very nature, be harder to locate and are likely to have more concerns regarding privacy.

It was not possible to find an estate purchased by an absentee landowner with prominent economic motivations, perhaps because an upland estate is unlikely to offer high returns to a financially motivated investor. Two estates were included in the 'Economic/Resident/Inherited' category (Estate 4 and Estate 9) because they offered

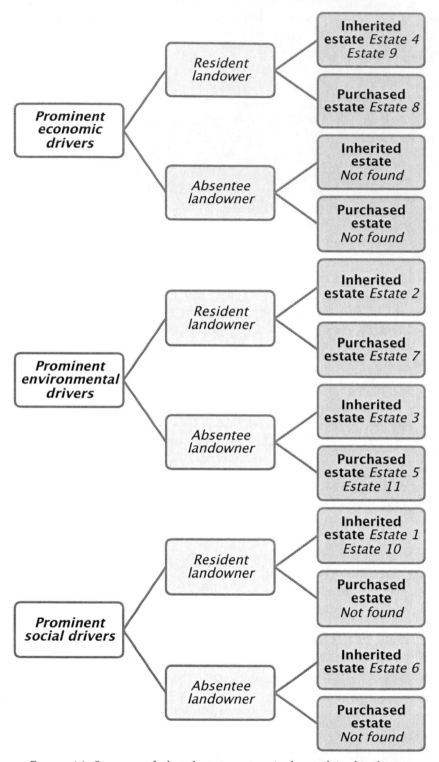

FIGURE 4.1 *Summary of selected estates, categorised as explained in the text*

a contrast in estate size:[2] Estate 4 was less than 10,000 acres (4,046 hectares), while Estate 9 was more than 50,000 acres (20,234 hectares). It was also not possible to find an estate which had been purchased by someone with predominant social drivers. Estate 1 and Estate 10 were both assigned to the 'Social/Resident/Inherited' category as they also offered a contrast in estate size: Estate 1 was small (less than 10,000 acres), while Estate 10 was large (more than 50,000 acres).

The eleven selected estates cumulatively covered an area of 142,516 hectares (352,164 acres). Their average size was 12,956 hectares (32,015 acres). Geographically, the estates spanned most of upland Scotland, including one on the north coast, two near the west coast, and one in the Southern Uplands, with the remainder being more centrally based around the Cairngorms National Park, in Aberdeenshire, Inverness-shire and Perthshire. The smaller estates with resident landowners were mainly farmed in-hand with a mix of sheep and cattle though one small estate and one medium-sized estate had given up farming on their own account owing to insufficient returns. Others, in contrast, were experts in their field and had concentrated on stock improvement as a means of increasing income.

Absentee landowners on estates of all sizes had either entered into contract farming arrangements or continued with existing agricultural tenancies. Two estates had a history of fish farming, though neither was actively involved at the time of the interviews. Universally, one of the main sources of estate income was property rentals, including commercial lets, residential property and holiday lettings.

All the estates considered let stalking and shooting as important economic activities on the estate even when economic considerations were not of primary importance. Forests, though present on most estates, were often managed for sport and amenity rather than commercially, particularly on the smaller estates. Some of the larger estates had a much broader range of commercial activities, including the management of historic buildings and related business ventures, such as film work, tourist activities, catering businesses, golf courses, and various livestock enterprises.

Each owner was asked to participate in an exercise that required them to arrange in order of priority a series of thirty-four statements about the strategic aims and objectives for the management of the estate, their understanding of sustainable management, and constraints to sustainable management of the estate. The method used for this exercise is called 'Q methodology' and is explained briefly in Box 4.1.

The statements used in the exercise were developed from the comments collected from the landowner surveys in conjunction with the study of the relevant literature to ensure that the statements were relevant and applicable. Each statement was printed on a card (four cards only had photographs: see Box 4.2). Each landowner was asked to arrange the thirty-four cards on a grid. The grid asked the question: 'In your vision for the future of your estate, which factors are most/least important?' The landowner placed each card at the relevant place along the grid towards 'most' or 'least' important, depending on his/her opinion, as shown in the photograph in Figure 4.2. The landowners were asked to consider each statement in the context of their own lifetime or tenure (landowners were between forty-five and seventy years old at the time of completing the Q-sort).

Box 4.1 Q methodology: an introduction

Q methodology was the innovation of William Stephenson (1902–89) in the mid 1930s and has been widely adopted as a useful tool in the field of social sciences (Brown 1980, 2005). A form of factor analysis, Q methodology focuses on the in-depth study of a few cases rather than superficial analysis of large numbers of subjects (Brown 1996). Specifically, the method can be used systematically to identify the range of distinctive subjective standpoints in any given context, using statistical inference (Stephenson 1975). For example, it allows in-depth exploration of people's perceptions and understandings, allowing these to be categorised into 'clusters' of opinions that can be analysed within a larger population. In particular, statistical analysis of the responses can establish: what patterns exist in the data and how they overlap; what characteristics are involved in what pattern and to what degree; and what characteristics are involved in more than one pattern.

In recent years, Q methodology has increased in popularity as the introduction of sophisticated computer software to carry out the complex statistical analysis has made its application less cumbersome.

The results of each Q-sort were analysed using computer software designed to identify patterns emerging from the data. In addition to the Q-sort exercise, an in-depth interview was conducted with each owner (the topics covered in the discussion are summarised in Box 4.3). Visits were also made to relevant areas on each estate to gain a deeper understanding of management practices and the various enterprises. In addition to facilitating a wider understanding of the estate activities, this discussion is an essential part of the Q-sort process as it provides insight into the thought processes of the landowner and enables the researcher to interpret the Q-sort results in an informed way.

The priorities and motivations of private landowners

The mathematical process of cross-comparison of the results of all the landowners carried out by the software produced various reports from which three main landowner profiles were built up and interpreted. Each of the profiles 1, 2 and 3 shown in Table 4.1 effectively represents a group of landowners who hold similar opinions, based on the correlations that occur within the sample group of landowners across the thirty-four statements in Box 4.2. In Table 4.1, the most important and least important statements for each profile are provided, and a wider interpretation is indicated by the motivational drivers based on a combination of the Q-sort data and the interviews. The final column also indicates which landowners correlated most highly with each 'profile'.

The statements on which there was most agreement were: 'Reduced bureaucracy', which scored strongly positively for all three profiles; and 'Reducing grazing pressure allowing greater biodiversity' and 'More conservation work', which both scored negatively for all three profiles.

> **Box 4.2** List of 34 Q statements
>
> 1. Greater use of renewable energy
> 2. Increased bird numbers and diversity
> 3. Increased native mixed woodland
> 4. Reduced grazing pressure allowing greater biodiversity of flora and fauna
> 5. Increased use of designations for environmental protection
> 6. More conservation work
> 7. Fewer 'blots on the landscape'
> 8. Re-wilding of native forest
> 9. Expansion of commercial forestry
> 10. Increased wildlife
> 11. Financial viability without subsidies
> 12. An improved grant system
> 13. Improved farming prices
> 14. Non-estate income to provide support
> 15. Greater diversification
> 16. Greater financial support for preservation of historical built heritage
> 17. More tourism-related activities
> 18. Reduced bureaucracy
> 19. Reduced capital taxes to enable the next generation to inherit without undue financial burdens
> 20. Photograph – communication links (train, ferry, post bus)
> 21. Improvements in infrastructure
> 22. More affordable housing
> 23. Increased rural industry
> 24. Greater links between the estate and the local community
> 25. Reform of planning rules allowing increased development
> 26. More job opportunities
> 27. Greater involvement with local businesses
> 28. More opportunities for the public to enjoy the benefits of the estate
> 29. Increased eco-housing
> 30. Increased recreational opportunities
> 31. Increased sporting opportunities
> 32. Photograph – black grouse/deer
> 33. Photograph – pheasant/grouse shooting/deer stalking
> 34. Photograph – crofting/small farmers

The Q-sorts indicate that the landowners in this sample place a considerable emphasis on the economic aspects of estate management. This is reinforced by comments made during the interviews such as: 'I don't care what interpretation environmentalists put on the word sustainable, most things at the end of the day come down to money.' Landowners also consistently highlighted the importance of making a

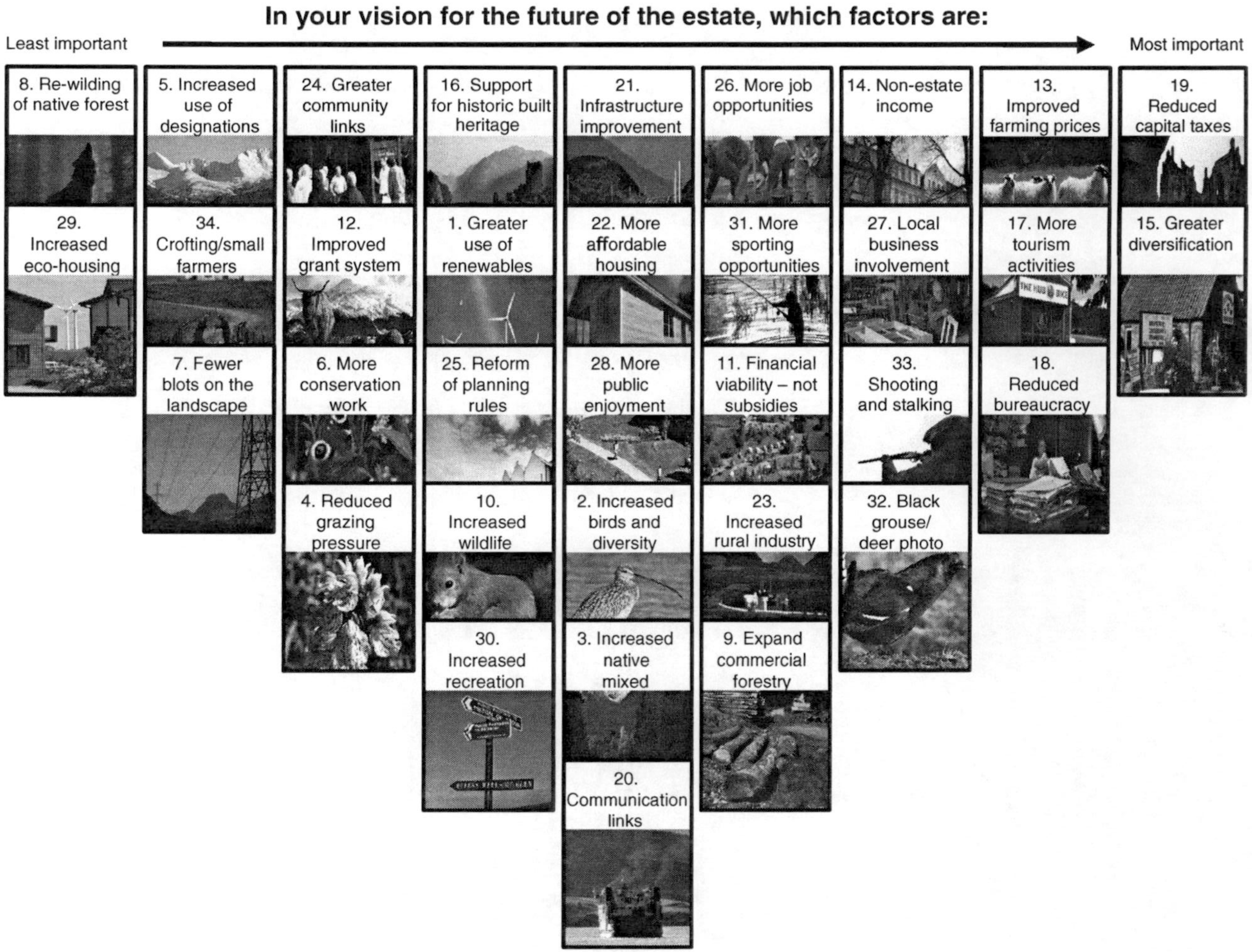

In your vision for the future of the estate, which factors are:

Least important → Most important

FIGURE 4.2 *Picture of Q-sort grid*

Box 4.3 Topics of discussion with the landowners

- **Fiscal (taxation):** establishing which taxes are, or could be, a consideration on the estate (e.g., Income tax; Capital Gains Tax; Stamp Duty; Inheritance Tax; VAT). Mechanisms in place or planned to mitigate taxes were considered. Questions were asked to establish the landowner's views on current legislation.
- **Agricultural and forestry:** exploring the extent to which the estate claims grants and subsidies (e.g. Single Farm Payment; Less Favoured Area payments; forestry grants; conservation grants, etc.). Understanding views on the decoupling of the Common Agricultural Policy (CAP) and the current Scotland Rural Development Plan (SRDP).
- **Environment and conservation:** considering whether the cross-compliance requirements for farmers are set at the right level, whether more conservation work would be carried out if the grants were less competitive, and what types of grants incentivise the landowner to take action.
- **Planning:** asking whether any planning applications had been made in recent years and how successful the process had been. Establishing views on current planning rules.
- **Social:** understanding changes in the local population, the impact of land reform legislation, local social infrastructure, rural industry, housing, and local employment opportunities.
- **Other:** understanding deer management, contact with neighbours, attitudes to hunting and sporting management, renewable energy options (including existing proposals for wind, hydro or biomass projects).

---

positive financial return: 'Does it [the estate] pay for itself?' 'Yes, it does. If it didn't, we would probably all be on the dole!' In a more long-term perspective, the results of the private landowner survey indicate that over 90 per cent of these landowners wish to pass on their estates to their heirs, and that no landowners want to sell, suggesting a clear underpinning economic motivation (see Chapter 3). This is supported by the correlations for Profiles 1 and 3 from the Q-sort results that place economic motives above others.

Investigating estate economics

Table 4.2 shows the economic viability of the selected estates, compared with the motivational drivers and main sources of income. Significantly, those estates that are economically driven (Estates 4, 8 and 9) are economically viable as a result of the earnings of business enterprises on the estate. The viability of these estates depends on the success of these businesses, whether farming or other diversified activities. For each of these estates, a resident landowner is the driving force behind each business (see Box 4.4).

Estates 1, 3, 4 and 7 are all small estates that correlate with Q-sort Profile 1

TABLE 4.1 Summary of motivational drivers, organised by profile

Profile	Most important statements	Least important statements	Motivational drivers
1.	Reduced capital taxes Improved farming prices Reduced bureaucracy Financial viability without subsidies Non-estate income to provide support Increased sporting opportunities	Re-wilding of native forest Crofting Increased use of designations for environmental protection Increased eco-housing Fewer blots on the landscape	Predominantly economic without government support, least interested in environmental issues **Estates 1,3,4,7,9 and 10**
2.	Increased rural industry More affordable housing More job opportunities Improved farming prices Reduced bureaucracy	Shooting Fewer blots on the landscape Increased wildlife Black grouse/deer photograph Increased sporting opportunities More conservation work	Predominantly social, least interested in environmental issues. **Estates 2 and negative 5ᵃ** (for which the results should be read in reverse)
3.	Improved grant system Reduced bureaucracy Non-estate income to provide financial support Shooting Greater use of renewable energy Reform of the planning rules allowing increased development	Improvements in infrastructure Community services photograph Re-wilding of native forest Greater involvement with local business More job opportunities Improved farming prices	Predominantly economic with government support, least interested in social issues **Estates 6 and 8**

ᵃNegative 5 means that the landowner of Estate 5 showed a negative correlation with Profile 2, indicating a profile completely the reverse of that of the landowner of Estate 2 and requiring the most and least important statements to be viewed in reverse, exhibiting a predominant motivational driver of environmental issues, least interested in social issues.

(demonstrating predominantly economic drivers, not seeking government support and least interested in environmental issues). These estates include in-hand farming enterprises, and therefore their owners attach high priority to 'improved farming prices'. The owners of the largest estates (Estates 9 and 10) also view this as a priority, though there is less dependence on in-hand farming; these estates have significant farming tenancies, for which rental income depends on agricultural returns.

In the longer term, some landowners expressed concerns about their ability to sustain the estate financially from one year to the next. One commented pessimistically: 'I think in the long term we are a bit of a dead duck.' Only just over a quarter of the estates surveyed in Chapter 3 survive without private external financial support from year to year and, even with this external 'subsidy', most are just breaking even and none of those would survive without public subsidies, suggesting that long-term economic sustainability is, at best, tenuous. The alternatives appear limited, however: landowners are faced with a choice of selling assets, borrowing money, or allowing the fabric of the estate to fall into disrepair: 'Without selling a chunk of the estate, in order to do something with those [derelict] cottages, we can't afford it.' The private

TABLE 4.2 Summary of the characteristics of selected estates

Estate	Size	Landowner resident or absentee	Estate purchased or inherited	Dominant motivational driver indicated by survey	Economically viable (with subsidies if claimed)	Q-sort profile	Economic viability depends on
1	Small	Resident	Inherited	Social	No	1	Subsidies and private wealth
2	Medium	Resident	Inherited	Environmental	Yes but does not support landowner	2	Diversified activities
3	Small	Absentee	Inherited	Environmental	Yes but does not support landowner	1	Private wealth to cover capital expenditure
4	Small	Resident	Inherited	Economic	Yes and supports landowner	1	Farming and subsidies
5	Medium	Absentee	Purchased	Environmental	No	2 neg.	Private income and wealth
6	Large	Absentee	Inherited	Social	Yes but does not support landowner	3	Subsidies equal profit
7	Small	Resident	Purchased	Environmental	Yes but does not support landowner	1	Renewable energy
8	Small	Resident	Purchased	Economic	Yes and supports landowner	3	Diversified activities
9	Large	Resident	Inherited	Economic	Yes and supports landowner	1	Diversified activities and property income
10	Large	Resident	Inherited	Social	Yes	1	Property income
11	Medium	Absentee	Purchased	Environmental	No		Private income and wealth

Box 4.4 The resident landowner being the driving force behind a business

A lifetime's dedication to breed improvement and promoting native breeds had enabled one landowner to achieve exceptionally high prices for his cattle. Through this dedication, the landowner had been able to retain his estate and to remain resident without relying on subsidies. His priorities were for the economic sustainability of the estate through farming and the ability to be able to hand over the estate to his heirs when the time was right. After all the work he had put in during his lifetime to ensure the continuation of the estate, his main concern was to avoid the estate having to be sold on his death to pay Inheritance Tax. Though this landowner showed little interest in environmental schemes, his dedication to high-quality traditional farming ensured that the land and landscape were retained in excellent condition.

landowner survey indicates that 39 per cent of the respondents have sold land for building purposes, though it is not known how many of those used the proceeds to pay for estate expenses. All the landowners acknowledged the requirement for frequent injections of capital to cover particularly bad years and to facilitate capital expenditure.

Landowners do not usually view their estate as a source of income and, in some cases, it is considered a burden when bequeathed without additional assets that could generate income: 'I think the logic is that it has worked for a long time already so let's see if it can continue, but if you want to privately educate three or four children you are stuffed really!' Though higher returns may be achievable from other ventures, case-study landowners typically report that grouse shooting is the major income source, both directly and indirectly from the rental of hunting lodges; for example: 'No, I am not remotely interested in shooting. It is not for me a priority, but as a landowner on this estate it is important. From a commercial point of view, from our point of view, grouse and deer are incredibly important.' (See Box 4.5 for an example.)

Box 4.5 The economic importance of grouse shooting

The most successful grouse moors are managed by teams of gamekeepers who aim to maximise the grouse population by minimising the threats of habitat loss, predation and disease. On one of the largest estates in the study, a single successful day of driven grouse shooting by the traditional nine guns ensures a year's employment for the gamekeepers. The emphasis on the economic motive behind grouse shooting on this large estate is crucial to the continued employment of gamekeepers who are also responsible for all the conservation work carried out on the estate. In addition, the shooting guests support the local community through spending in local shops, restaurants and hotels.

With farming revenues at a low ebb, the economics of marginal upland agricultural holdings are not seen to provide an alternative to grouse shooting: 'We are probably on the brink of getting rid of our sheep, Scottish blackface. We have got a breeding flock of 630 and that is not enough to sustain a shepherd.' Further evidence that multiple drivers influence decision-making was provided by the widely shared desire to continue sheep farming, often to help control ticks and to claim agricultural subsidies rather than for the (generally limited) agricultural income that would be generated. The need to reduce agricultural rents to address the tick problem is a reflection of the poor market price and low margins available to sheep farmers. Owners of Estates 1, 3, 4, 7, 9 and 10 all correlated with Profile 1 in the Q-sort (see Table 4.1) in which 'improved farming prices' were viewed as being most important for the future sustainability of the estate, alongside 'financial viability without subsidies'. The landowner interviews provide evidence that, while agriculture may not in itself be an economic driver, it works as a facilitator for grouse management and other activities (Box 4.6).

Box 4.6 Agriculture as a 'facilitator'

The landowner of one of the largest estates in this study stressed the need for income to cover expenditure. On this mixed estate, the agricultural income arising and resources available from ownership of the higher-grade land allowed for continued support of farming activities at an appropriate level on lower-grade, unprofitable land which he considered to be beneficial both environmentally and socially. The goal of retaining the whole estate intact to pass from generation to generation allowed the landowner to take an holistic approach. While being conscious of the overriding need for economic sustainability, this landowner had recourse to resources that would not be available to the smaller upland landowner if the land were to be fragmented.

Some landowners were searching for alternative, more lucrative sources of income: 'I would imagine by two years' time we will probably be back to breaking even . . . by diversifying into the tourist industry.' There was, however, reluctance from others to exploit these more 'modern' income streams. Armstrong's (1980) suggestion that the pursuit of profit may be considered distasteful was still valid three decades on; the old-fashioned notion that diversifying into trade is socially unacceptable was still held by some of the landowners. Even for those who were willing to consider it, diversification was not necessarily seen as a panacea (Box 4.7).

In the long term, many of the landowners are concerned about the implications of inheritance tax: though most accept in principle that 'death duties' are inevitable, some feel that the current system is neither equitable nor affordable. Value Added Tax (VAT) is also regarded as a challenge, in particular with regard to the limitations on claiming input VAT on expenses relating to rented property. Many regard the rules as a disincentive to investment in existing housing stock, thereby limiting housing availability in rural communities. There were several calls by landowners for rural residential let property to be zero rated to solve the problem of it being considerably

Box 4.7 Estate business diversification

A small estate in an ideal position to take advantage of income from summer tourism had diversified its activities to include the operation of a fish restaurant. The process was not without difficulty, with bureaucracy being cited as the biggest disincentive to investment. The income from the business helped to support one family member, reducing the burden on the rest of the estate and allowing that individual to remain in the area. Being seasonal, however, and providing only modest returns, the business was not a solution to the problems faced by the estate as a whole.

more expensive to repair old buildings than to build new ones (which *are* zero rated for VAT).

Landowners were also asked for their views on grants and subsidies, particularly in view of the Scottish Government's vision of a market-based agricultural sector delivering public goods under a light-touch regulatory regime. Though landowners agree in principle with the concept of the decoupling of the CAP, moving away from production subsidies, they are unable to foresee a future for farming in the uplands without some form of financial support. Landowners believe that they should be rewarded for providing public goods, and that grants for conservation work should continue to be the mechanism through which this is delivered. They criticise the competitive nature of the schemes and the lack of transparency, however. There is also resentment that, under some grant scheme options, landowners are reimbursed only when they produce receipts to show that they have paid others to complete the work, rather than being paid for doing the work themselves, as this does not help their 'bottom line'. For example, under the Land Managers' Options (2008–2013), creation of access for sustainable forest management attracts a grant payment of just 50 per cent of the amount paid to a contractor, resulting in a net cost to the landowner of 50 per cent. For landowners without sufficient funds (most of whom are economically motivated), there is no financial incentive to do such work. Landowners feel that the regulations associated with the Scotland Rural Development Programme (SRDP) are confusing and ineffective, exacerbated by inefficient implementation and bureaucracy.

Overall, economic motivation emerges most strongly where the landowner is resident on the estate and plays an active part in the day-to-day management of estate business(es). Their presence on the estate is often linked with increased entrepreneurship, diversified activities and more productive, high-value farming. Landowners in this category have a tendency to view 'sustainability' as a short-term, monetary issue. Grouse shooting is often the primary source of income but, because of doubts about its future, diversified activities and control of financial risk are also priorities. The biggest threat to sustainability among this group of landowners is a lack of funds to cover large items of expenditure, fund repairs, or pay inheritance tax. A lack of funds also hinders consideration of environmental and social issues beyond those funded by grant income or incidental to other activities such as grouse shooting.

The estate's natural heritage

Caring for the natural heritage is an important focus only on estates that have a financial 'cushion' of some kind (see Table 4.1). The owners of Estates 2, 3, 5, 7 and 11 all have other sources of income. Despite this, the results of the Q-sort indicate that these estates all have high primary economic motivation; the environmental aspects of estate management are important, but regarded as a secondary motive by most.

In the Q-sort, only the owner of Estate 5 ranks environmental concerns above all others, correlating negatively with Profile 2 (indicating strong environmental motives). Unfortunately, the owner of Estate 11 was unable to complete the Q-sort but interviews with three managers and advisers working for the estate suggest that caring for the natural environment of the estate is also a primary motivation for this landowner (see Box 4.8). It is worth noting that the owner of Estate 11 is the only female within the sample; Stern et al. (1993) found that women tend to take a more pro-environmental stance and give a higher priority to altruism. A key distinguishing factor of Estates 5 and 11 is that the owners have recourse to sufficient means to support their estates from external sources.

Box 4.8 Environmental priorities

Without the economic restrictions that hinder many upland Scottish estates, the highly motivated landowner of an estate exhibiting strong environmental priorities had a formal written five-year plan setting out the various goals to be achieved. Expert advice had been sought from various specialist individuals and nature conservation organisations, and work to restore, improve or create habitats for wildlife was the main theme of the plan. Since purchasing the estate, this landowner had drastically reduced sheep and deer numbers to prevent overgrazing, particularly of riparian areas, fenced large tracts of young Caledonian pine to allow natural regeneration, and improved poorly constructed access tracks that were causing soil erosion. The agenda was not driven by grants and subsidies – none was claimed.

Seven of the landowners ranked the environmental statements in the Q-sort exercise as least important. The responses to the broader survey (see Chapter 3) show that environmental motivation is strongest where the estate does not financially support the landowner, and particularly for absentee landowners who purchased their estates rather than inheriting them. For these landowners, landscape quality is a top priority. Visits to the estates showed that, in most cases, actions could be taken to improve visual aspects of the landscape though some were beyond the scope of the landowner. Two examples are as follows. On one estate, there were several miles of redundant electricity pylons that had not been removed following the creation of a new line. Many of the smaller issues were being addressed with the resources that were available, however. On another estate, the landowner explained that his staff constantly had to deal with incidents of fly-tipping on the land, clean up after campers, or remove blown plastic bags from roadside fencing. All of the landowners were keen to demonstrate that their estates maintain a diverse and thriving wildlife habitat but most acknowledged that further work was dependent on funding.

New absentee landowners not dependent on their estates for financial support are more likely than established landowners to be motivated by environmental objectives but none of those interviewed held strongly pro-environmental views, with most believing that turning the Highlands 'back into a wilderness' would be akin to excluding human activity. Like most farmed and managed landscapes, the Highlands support a community of agricultural workers, gamekeepers and other employees, ensuring the retention of local services in remote areas. One landowner who had purchased his estate commented: 'The problem with a strong environmental agenda is that it would destroy the local community.' Those living on the estate of the most environmentally motivated landowner were wholly reliant on the external funding provided by that landowner; no agriculture, little game and no other diversified enterprises meant that the estate provided no local economic opportunities. Without continued external financial support from the landowner, the community would disappear because the estate itself is not economically viable.

In terms of environmental policy, and landowners' views on habitat and species protection, there are widespread concerns that the current grant system does not recognise the full extent of the costs involved in providing public benefits. On the case-study estates, however, many land-use practices were carried out with consideration for the environment, and landowners believe strongly that their gamekeepers are a primary means of ensuring the continued survival of many species. For example, landowners view the practice of muirburn as beneficial for other ground-nesting birds, not just the grouse that are the gamekeepers' priority. Another example is the riparian growth, that landowners encourage because it is beneficial for salmon stocks, also providing improved habitat for invertebrates and mammals such as otters.

The results show that all landowners are keen to facilitate some pro-environmental work, and that this sentiment is strongest among those who purchased their estates and do not need to rely on estate income. This has led to investment in habitat quality, aesthetic improvements in the landscape, and reduced levels of pollution. Where environmental motives are combined with economic motives, there is a bias towards schemes that improve habitat for grouse and salmon, with a recognition that these offer considerable additional benefits for other species. Heavy investment in keepering, funded by grouse shooting, is seen as environmentally beneficial overall. Grant income is particularly helpful where conservation and sporting objectives coincide (for example, reducing deer and hence tick numbers). The restrictions on predator control, when these species reach levels where they cause substantial damage to other wildlife, are seen as a constraint by some. Some landowners fear that a single-minded focus on environmental protection could stifle economic development opportunities, thereby undermining the social and economic sustainability of local communities.

The social implications of private landownership

The owners of Estates 1, 6 and 10 appear to be primarily motivated by social considerations (Table 4.2). All these estates were inherited, and their owners have alternative sources of income so are not dependent on the estate providing financial income. The strongest social motivations exist on estates whose owners have built links with communities over many generations and the landowner's family has a history of involvement with locals and visitors (Ilbery 1983; Gasson and Errington 1993). This is particularly evident on Estates 6 and 10 (both of which are large estates) and, to some extent, on Estates 1 and 2. Many owners of upland estates have traditionally adopted a paternalistic attitude to decision-making (Stewart et al. 2001), and the results show that this persists among some socially motivated landowners today. For example, one commented: 'It would have been more profitable not to have any sheep. I thought about it, but that would remove somebody's livelihood.'

The economic benefits of grouse shooting for local communities have been highlighted in recent studies. A survey conducted by The Fraser of Allander Institute in 2010 estimated that grouse shooting accounted for 46 per cent of the permanent employment across the 304 upland estates they surveyed. Estate 6, the largest of those studied in detail, makes only a very modest profit each year (almost equal to the subsidies received), and this is kept in reserve to cover future shortfalls or help fund capital projects:

> We have fifty or sixty people involved on each day's grouse shooting and a lot of those are being paid by us. That cash is going straight into the local community. You can see from the figures that it isn't the 'rich' landowners making the money. The money from grouse shooting stays here. It all goes back in the local economy.

Most estates try to support the local community through employment opportunities, though this is not always practicable. On some estates, new money and the entrepreneurial spirit that Kalantaridis (2010) identified as a catalyst have considerably increased levels of employment. In addition, landowners often take it upon themselves to address issues normally left to the public sector (Beedell et al. 2009). It is sometimes difficult to distinguish, however, whether the development of housing is motivated by financial gain or genuinely done to support the local community. For example, the owner of Estate 2 is the only one to correlate positively with Profile 2 in the Q-sort, suggesting that the landowner has strong social motivation. This landowner has considerable interest in developing residential property in the area but no interest in grouse shooting. His interests and motivations are strongly influenced by the fact that the future economic viability of his estate is dependent on residential and commercial property development. Though his motives appear to be socially oriented, he is also likely to be driven by economics. Socially motivated landowners have a firm understanding of the link between successful tenants and a healthy rent roll. Their commitment to the community helps to maintain a 'critical mass' to sustain community functions such as schools and transport. Relationships are seen as two way, with communities benefiting from wealthy visitors to high-quality sporting estates. Partnership working provides impetus for social cohesion and resilience.

Most estate housing is not exploited to the full – rents are not maximised, often a result of paternalism, but also housing is 'affordable' by default, usually as a result of lack of investment. Properties let on long-term leases command particularly low rents because of the poor condition of the fabric of the property and the lack of investment in modern boilers, insulation, and so on. So, the provision of new 'affordable' housing may be a by-product of a difficult economic situation rather than being socially motivated landownership (see Box 4.9). In most areas, holiday accommodation offers the most lucrative source of income but requires a high level of investment to achieve the best returns. Though holiday accommodation reduces the amount of housing stock available for the local community, there are mixed feelings among the landowners in this regard because encouraging visitors is seen as a positive contribution to the local economy. Though most landowners regard tourism as beneficial, Frew and Hay (2011) point out that both real returns from the tourist industry and visitor numbers are flat or declining in Scotland, and that most of those visitors are from Scotland itself so are merely causing a displacement of income within the country. Nevertheless, it is economically beneficial to the Scottish economy, as well as to the local economy and the individual estate, for Scottish residents to spend their holidays in Scotland rather than elsewhere, and the increase in facilities for them may encourage more 'staycations'. Tourism is considered to be central to the development of many small communities but Beeton (2010) suggests that there tends to be a limited understanding of the complex nature of tourism and its relationship with community development. To succeed, a community requires not only individual entrepreneurial landowners or tenants to develop businesses but also a community-based entrepreneurial approach to realise the potential fully.

Box 4.9 Housing and rental income

Part of the rental income of one small estate included that from a local hotel. Currently let on a fifteen-year tenancy, the landowner was not confident that the current tenant would continue because there had been ten different tenants in the last twenty years. Despite the area being popular with walkers, the hotel does not appear to be a particularly successful business. The landowner explained that there was very little in the area to interest tourists. The estate had only one cottage and this was let to a 'social' tenant. Despite the beauty of the area's landscape, the landowner was of the opinion that holiday lettings would not be a viable alternative, even if he could terminate the tenancy.

On one of the estates, the paucity of economic resources has resulted in the lease of an historic building at a peppercorn rent for fifty years, giving use and responsibility to a community organisation. As on many estates visited, the maintenance of historical buildings is a major drain on the finances of this estate. The social motivation behind such a bold gesture is secondary to the economic reality that the estate would be unable to fund the ongoing maintenance of the building. Consequently, the community offers the best financial solution to the problem, removing the landowner's social obligation to maintain and manage the building for the benefit of the community and the wider public interest while retaining long-term ownership in the interests of the estate. This is an example of a formal structure for community integration but informal alliances have also been developed for the mutual benefit of the community and this estate, achieved with limited economic resources or used as a method of securing long-term sustainability without the need for the estate to provide short-term cash injections.

Social cohesion, which holds the individuals in a rural area together in an informal support network, is viewed as important by most of the interviewed landowners who recognised the benefits of living in a trusted and supportive community. As one landowner tried to explain, however, not all incomers understand the need for co-operation and mutual assistance. In areas with limited resources for responding to unexpected problems, good neighbours are seen as vital. On one remote estate, the landowner recounted the story of a neighbour who had been hospitalised during a particularly harsh winter. The sick landowner was amused by the fact that several of his neighbours had kindly fed the sheep for him – each unaware that they had already been fed by someone else. Another landowner pointed out the implications of bringing in labour from further afield: 'If you regularly use a plumber from Inverness, the local man is unlikely to be willing to turn out on a cold winter night in an emergency.'

The landowner of Estate 2, who correlated strongly with the social motivations in the Q-sort, is influenced by the estate's strong financial dependence on income from commercial ventures within the community. Its links with that community are very strong, particularly as the current estate was part of a larger estate that has been broken up over time, so that generations of different branches of the family live within a relatively small area and many are involved in the various activities that take place on, or close to, the estate. On a couple of other estates, the owners do not consider themselves to be part of a local community but most acknowledge that the estate is an integral part of a community even if they regard the term 'local' as inappropriate. Though the owners of most of the estates show some concern for the local community, those owning the smaller estates, particularly those that had been purchased more recently, display less interest in integration. This may partly be because the smaller the estate is, the less likely it will have a settlement within its boundaries; equally, where a community is bordered by several estates, allegiances to any one estate may not be as strong.

Regarding social policies, several landowners expressed particular concerns that planning regulations constrain development in rural areas. Landowners understand the need for planning controls but are continually disappointed by the lack of flexibility and forward thinking shown by local planning officers, one landowner referring to his local planning officer as 'Stop-the-job Bob'. The law regarding tenancies and tenant rights is also an issue of concern. Most landowners are reluctant to issue new tenancies, preferring to enter into farming contracts or farming in-hand, because they find the legislation particularly 'draconian'.

The landowners highlighted aspects of Scottish outdoor policy that conflict with the protection of the landscape and wildlife. Though none of the landowners interviewed spoke negatively about public access to estate land, many feel that they are entitled to some form of compensation for managing access facilities such as paths (as considered in Chapter 2). On the estates visited, there are examples of recreational activities and of camping that have caused environmental damage but many of the landowners have tolerated this and taken no action (see also Box 4.10). With no obvious avenue of recourse, there is a lack of awareness of how to address such breaches of the public right of responsible access. Legal action is regarded as too expensive, with causation and loss difficult to prove and the outcome not likely to bring economic benefit, as compensation would be limited to the financial loss incurred.

Box 4.10 Problems with campers

An estate manager in a particularly popular area for camping explained that, although the estate was very tolerant of individual responsible camping, it 'drew the line' at camping parties. Without any formal cooking, sanitation or waste facilities, and attracting young campers who were often ill-prepared, these 'unorganised' parties meant that the environmental damage caused was often considerable, and the job of cleaning up the site was a major task for estate staff. The manager explained that he had now started patrolling vulnerable roadside areas during the 'party' season to 'move on' campers before the worst of the damage occurred.

THE IMPACT OF MOTIVATION ON LANDOWNER DECISIONS

In summary, the research on the eleven selected estates considered the motivations of landowners with regard to economic, environmental and social activities and/or challenges. A key finding, represented in Figure 4.3, is that when economic drivers are satisfied, landowners tend to focus more on the natural heritage and social aspects of estate management decisions, the latter often engendered through long-term connections with the community. Crucially, this demonstrates the importance of these two factors in the sustainability of upland estates, both of which are essential in achieving balance. An evenly balanced approach was found only on two of the largest estates in the sample, as all other estates lacked at least one of these two factors, time or money.

FIGURE 4.3 *Summary of landowner motivation in the context of money and time*

TABLE 4.3 Summary of positive and negative outcomes of economic, environmental and social landowner motivation

Landowner motivation	Positive and negative outcomes found on selected estates	
Economic	**Positive**	Leads to an increase in entrepreneurship, diversified business activities, and more productive farming.
	Negative	Lack of funds can hinder consideration of social issues and short-term gains from development. Asset exploitation may be detrimental to the environment.
Environmental	**Positive**	Increase in quality of habitats and species diversification provided conservation work is not overly species-selective. Aesthetic improvements in landscape and reduced pollution.
	Negative	Protection causes intended or unintended exclusion of people as a lack of commercial development or housing can cause breakdown of community and a spiral of decline.
Social	**Positive**	Partnership working provides impetus for social cohesion and resilience. Estates can provide the core of critical mass needed to sustain community functions such as schools, post offices and transport links.
	Negative	Expansion of employment opportunities and housing can have a negative impact on landscape and ecology.

On the eleven estates considered in this chapter, the environmental and social factors were generally less important to the landowners, as retention of the estate and day-to-day financial survival limited the scope for other considerations. The landowners who correlated most strongly with economic motivations in both the survey and the Q-sort were also personally active in maximising potential income through either high-value farming or diversified estate activities. On all these estates, the owners were aware that owning an estate was not the most financially efficient use of their capital and, therefore, intrinsic motivation, valuing the ownership or management of the estate as an activity in its own right, tended to be a factor in the decision-making process.

This chapter has demonstrated the influential role of landowners' motivations in management decision-making. This, in turn, leads to different outcomes depending on what those motivations are. Table 4.3 provides a summary of some of the positive and negative effects of the different drivers demonstrated in the case studies.

This research has shown that, although appropriately designed government policy has the potential to influence some landowners' decision-making processes through financial incentives, many prefer not to rely on subsidies, and most environmental work is carried out by those who are financially independent. The results also indicate that resident and inherited landowners can contribute to the maintenance of resilient communities in the Scottish uplands. Such established landowners often have an important role in supporting the social sustainability of these rural areas. Of course, not all private landowners are similarly 'enlightened'; there are 'bad apples in every bunch', and instances of mismanagement and neglect do occur (and are widely reported). Nevertheless, given that privately owned estates are likely to remain a significant piece of the upland jigsaw for many decades to come, it is important to recognise their beneficial potential and to understand the motivations influencing land-management decisions, so that such estates can be enabled to play their part in contributing to the economic, social and environmental well-being of the Scottish uplands.

NOTES

1. Work with the owner of Estate 11 was not fully completed because, despite agreeing to take part, the owner remained elusive. Meetings were instead conducted with estate staff. The estate was still included as 'Environmental/Absentee/Purchased', though the data gathered from this estate were incomplete.
2. Estates were classified as 'small' (less than 10,000 acres or 4,046 hectares), 'medium' (10,001–50,000 acres or 4,047–20,234 hectares) and 'large' (over 50,000 acres or 20,234 hectares). Exact estate sizes are not given in order to preserve anonymity.

REFERENCES

Armstrong, A. M. (1980). *Geographical aspects of the ownership, management and use of rural land on landed estates in the northern Highlands.* PhD thesis, University of Aberdeen.

Beedell, J., Annibal, I., Teanby, A., Hindle, R. et al. (2009). 'Working positively with rural estates: the scale and nature of estates and their contribution to the East Midlands of England'. Report prepared for East Midlands Development Agency. May 2009. <www.emda.org.uk/rural estates/> (last accessed 29 August 2011).

Beeton, S. (2010). 'Regional community entrepreneurship through tourism: the case of Victoria's rail trails'. *International Journal of Innovation and Regional Development* 2 (1/2), pp. 128–47.

Brown, S. R. (1980). *Political subjectivity: applications of Q methodology in political science.* Yale University Press, New Haven. 355 pp.

Brown, S. R. (1996). 'Q Methodology and Qualitative Research'. *Qualitative Health Research* 6 (4), pp. 561–7.

Brown, S. R. (2005). 'Applying Q methodology to empowerment'. In: Narayan, D. (ed.), *Measuring empowerment: Cross-disciplinary perspectives.* The World Bank, Washington, DC, pp. 197–215.

Burton, R. J. F. and Paragahawewa, U. H. (2011). 'Creating culturally sustainable agri-environmental schemes'. *Journal of Rural Studies* 27 (1), pp. 95–104.

Denman, D. R. (1965). 'Land Ownership and the Attraction of Capital into Agriculture: A British Overview'. *Land Economics* 41 (3), pp. 209–16.

Fraser of Allander Institute (2010). 'An economic study of Scottish grouse moors: an update'. Game and Wildlife Conservation Trust, Perth.

Frew, A. J. and Hay, B. (2011). 'The development, rationale, organisation and future management of public sector tourism in Scotland'. *Fraser of Allander Economic Commentary* 34 (3), pp. 63–76.

Gasson, R. M. and Errington, A. (1993). *The Farm Family Business.* CAB International, Wallingford. 304 pp.

Gray, J. (1998). 'Family Farms in the Scottish borders: a practical definition by hill sheep farmers'. *Journal of Rural Studies* 14 (3), pp. 341–56.

Heley, J. (2010). 'The new squirearchy and emergent cultures of the new middle classes in rural areas'. *Journal of Rural Studies* 26 (4), pp. 321–31.

Hutson, J. (1987). 'Fathers and Sons: Family farms, family businesses and the farming industry'. *Sociology* 21, pp. 215–29.

Ilbery, B. W. (1983). 'Goals and values of hop farmers'. *Transactions of the Institute of British Geographers*: New Series 8 (3), pp. 329–41.

Kalantaridis, C. (2010). 'In-migration, entrepreneurship and rural–urban interdependencies: The case of East Cleveland, North East England'. *Journal of Rural Studies* 26 (4), pp. 418–27.

MacGregor, B. D. (1988). 'Owner motivation and land use on landed estates in the north-west Highlands of Scotland'. *Journal of Rural Studies* 4 (4), pp. 389–404.

MacGregor, B. D. and Stockdale, A. (1994). 'Land use change on Scottish highland estates'. *Journal of Rural Studies* 10 (3), pp. 301–9.

MacMillan, D. C., Leitch, K., Wightman, A. and Higgins, P. (2010). 'The management and role of Highland sporting estates in the early 21st Century: the owner's view of a unique but contested form of land use'. *Scottish Geographical Journal* 126 (1), pp. 24–40.

Mather, A. S., Hill, G. and Nijnik, M. (2006). 'Post-productivism and rural land use: cul de sac or challenge for theorization?' *Journal of Rural Studies* 22 (4), pp. 441–55.

Miller, D., Sutherland, L. A., Morrice, J., Aspinall, R., Barnes, A., Blackstock, K., Schwarz, G., Buchan, K., Donnelly, D., Hawes, C., McCrum, G., McKenzie, B., Matthews, K., Miller, D., Renwick, A., Smith, M., Squire, G. and Toma, L. (2009). 'Changing land use in rural Scotland – Drivers and decision making'. Rural Land Use Study Project 1. The Scottish Government, Edinburgh. < http://www.scotland.gov.uk/Publications/2010/01/06100615/0> (last accessed 12 December 2012).

Moxnes Jervell, A. (1999). 'Changing patterns of family farming and pluriactivity'. *Sociologia Ruralis* 39 (1), pp. 110–16.

Petrzelka, P., Malin, S. and Gentry, B. (2012). 'Absentee landowners and conservation programs: Mind the gap'. *Land Use Policy* 29 (1), pp. 220–3.

Riley, M. (2009). 'The next link in the chain: children, agri-cultural practices and the family farm'. *Children's Geographies* 7 (3), pp. 245–60.

Schneider, F., Ledermann, T., Fry, P. and Rist, S. (2010). 'Soil conservation in Swiss agriculture – Approaching abstract and symbolic meanings in farmers' life-worlds'. *Land Use Policy* 27 (2), pp. 332–9.

Stephenson, W. (1975). *The study of behaviour: Q-technique and its methodology*. University of Chicago Press, Chicago. 376 pp.

Stern, P. C., Dietz, T. and Kalof, L. (1993). 'Value orientations, gender and environmental concern'. *Environmental Behaviour* 25, pp. 322–48.

Stewart, R., Bechhofer, F. and McCrone, D. (2001). 'Keepers of the land: ideology and identities in the Scottish rural elite'. *Identities* 8 (3), pp. 381–409.

Veblen, T. (1908). *The theory of the leisure class: an economic study of institutions*. Macmillan, New York. 336 pp.

Wallace, M. (1998). 'Multiple Goals in Farm Family Decision Making: A Recursive Strategic Programming Analysis'. *Proceedings of the Agricultural Economics Society of Ireland* 1998/99, pp. 17–35.

CHAPTER FIVE

The laird and the community

Annie McKee

INTRODUCTION

The dominance of private landownership in upland Scotland raises questions about the impacts of the actions of landowners on socioeconomic development in upland areas (McIntosh et al. 1994; Wightman 1996, 1999; Shucksmith and Dargan 2006). Land-management decisions are based on external drivers, such as policy, funding opportunities and regulation (as outlined in Chapter 2) and, to a certain extent, on the individual motivations of the landowner (as discussed in Chapter 4). As a result, land-use decisions 'have a significant impact on potential for growth or decline in Scottish rural settlements' (Bird 1982, 55).

Recent research indicates that there is a lack of public understanding regarding contemporary private estate management (Fawcett and Costly 2010). Public perceptions of private landownership in Scotland often conjure up stereotypical images of the tweed-clad 'laird', as depicted in Compton MacKenzie's novel *Monarch of the Glen* (and its later popular television adaptation). Furthermore, private landowners are often criticised for negative influences on the sustainability of upland communities: for example, in terms of land management practices, estate developments (or lack of development), and the level of community involvement in estate management planning and decision-making (see Wightman 2010, for example). On the other hand, many long-established landowning families are known to have a strong sense of stewardship and a long-term commitment to the communities on their land, bringing in outside investment through private business and sporting lets that support the rural economy (Warren 1999; Samuel 2000; Higgins et al. 2002; Kerr 2004; Buccleuch 2005). The relationships between landowners and local communities are therefore contested and complex.

The Land Reform (Scotland) Act 2003 (LRSA) has led to the 'land question' becoming more community oriented (Bryden and Geisler 2007) (see Chapters 3 and 6); and the open access rights it includes challenge the owners of privately owned estates no longer to consider them 'private' spaces for their sole benefit. The Act also challenges landowners to increase 'community involvement in the way the land is owned and used' (Land Reform Policy Group 1999, 4). In the context of a changing rural population, with trends of counter-urbanisation, youth out-migration and declining land-based employment, questions arise with regard to who comprises the contemporary 'estate community', and what role (or roles) the private landowner should play in sustaining that community.

To date, there have been no thorough evaluations of interaction and engagement between private landowners and 'estate communities'. The importance of integrated community engagement, empowerment and 'bottom-up' initiatives for sustainable development cannot be overstated (Bridger and Luloff 1999; Fraser et al. 2006) and, as outlined in Chapter 1, there are also many opportunities for exploring participatory and collaborative approaches to land management.

To address this gap in our knowledge, this chapter presents the findings of research that studied six privately owned estates in depth to develop an understanding of the contemporary roles of the private landowner in facilitating sustainable 'estate communities' in upland Scotland. The research aim and objectives upon which this chapter is based are presented in Box 5.1. Through evaluating the interactions and engagement processes between estates and their local communities, the chapter considers the prospects for partnership approaches, and offers some best-practice recommendations for private landowners, rural communities and other agencies, as a contribution to promoting sustainability in the Scottish uplands.

Box 5.1 Research aim and objectives

The overall aim of this research was to explore the contemporary role of private landownership in facilitating sustainable rural communities in upland Scotland. Specific objectives therefore included:
- To examine the interactions between private landowners, estate representatives and rural communities;
- To evaluate the processes and practices of estate–community engagement;
- To explore potential estate–community partnerships, their strengths, weaknesses, opportunities and threats;
- To develop practical recommendations for ways in which owners of private upland estates can promote the sustainability of rural communities through good practice.

EXPLORING INTERACTIONS BETWEEN LANDOWNERS AND COMMUNITIES

Using the results of the private landowner survey presented in Chapter 3, six estates were identified as conducting 'proactive' community engagement practices. Information about these estates is provided in Table 5.1.

To understand interactions between the 'estate' and the 'estate community' in detail (see Box 5.2 for the definition of an 'estate community'), the researcher lived and worked on each estate, and within its local community, for several weeks. Where possible, she attended every community event that took place during her stay,

TABLE 5.1 Characteristics of the six estates,[a] including ownership structure (after Hutley 2011) and community size

Case study	Region	Ownership structure	Resident/ absentee landowner	Key enterprises	Community size[b]
1	Aberdeenshire	Trust	Resident	Residential and farm tenancies, guest house	570
2	Badenoch and Strathspey	Combination: Sole trader (private owner), trust and limited company	Resident	Forestry, in-hand farming, residential and business tenancies	620
3	Aberdeenshire	Sole trader (private owner)	Absentee	Sport shooting, holiday cottages	500
4	Argyll	Combination: trust, limited company and partnership	Resident	Renewable energy, holiday cottages	200
5	Argyll	Sole trader (private owner)	Resident	Renewable energy, residential and commercial tenancies (fish farms)	180
6	Sutherland	Sole trader (private owner)	Absentee	Crofting tenancies, fish farms	130

[a] Estate names are not stated to preserve anonymity.
[b] Approximate population established through pre-fieldwork desk-based study and discussion with research participants, as census data do not provide population detail at the scale of the individual community associated with an estate.

volunteered in a range of community businesses and activities (local pubs, village shops, cafés, community centres, community groups and fund-raising events), and carried out various jobs with estate maintenance staff, gamekeepers, gardeners and housekeepers. A research diary was used to record observations, informal conversations, thoughts and insights. On each estate, a household questionnaire survey was carried out, and in-depth interviews were conducted with a range of key community members and estate staff/tenants (see Box 5.3 and Figure 5.1). Figure 5.2 summarises the data collection and analysis process.

Box 5.2 Definition of 'estate community'

For this project, the 'estate community' was defined as the landowning family or organisation, estate employees (direct and indirect), and estate tenants (residential, agricultural and commercial). In some cases, the 'estate community' also included the population of adjacent villages, owner-occupiers within estate boundaries, estate visitors, and those with a 'sense of belonging' to the estate. This definition was established from the estate interviewees (see Box 5.3).

Box 5.3 Interviewees on the estates

Community Members: including local business people, community council members and other community group leaders.

Private landowners: including individual owners and directors of estate companies, as well as members of the owner's family.

Other estate management personnel: including estate manager/factor, and assistant estate manager/factor (both resident and agent).

FIGURE 5.1 *A walking interview with a crofter*

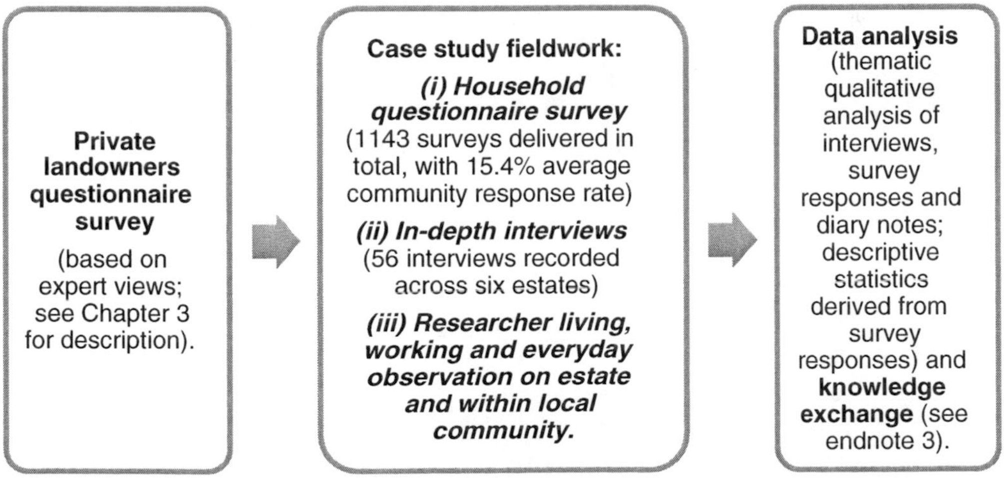

Private landowners questionnaire survey

(based on expert views; see Chapter 3 for description).

Case study fieldwork:

(i) Household questionnaire survey
(1143 surveys delivered in total, with 15.4% average community response rate)

(ii) In-depth interviews
(56 interviews recorded across six estates)

(iii) Researcher living, working and everyday observation on estate and within local community.

Data analysis
(thematic qualitative analysis of interviews, survey responses and diary notes; descriptive statistics derived from survey responses) and **knowledge exchange** (see endnote 3).

FIGURE 5.2 *Overview of data collection and analysis process*

IDENTIFYING KEY ISSUES

The results of the interviews, household surveys and fieldwork observations suggest that the influences and interactions of private landowners with their 'estate communities' centre around three important and interlinked issues: housing; employment; and community spirit or cohesion.

Housing

Many areas in upland Scotland lack affordable housing (see Thomson 2012) while, at the same time, in-migration is enlarging some communities. The challenge of providing affordable housing is evident on all six estates despite their disparate locations. On each estate, land-use decisions influence the provision of affordable housing; developments are constrained by planning restrictions, a lack of land availability (often due to estate objectives), high infrastructure costs (such as utility connections), and inflated land prices once planning permission has been granted. One central reason for the lack of affordable housing is a mismatch between typically low wages and an increasingly expensive rural housing market.

Affordability decreases as demands for rural housing increase. Relatively wealthy 'incomers' who move to remote rural areas from urban areas and higher-paid employment, often seeking a rural lifestyle while self-employed or for retirement, push housing prices above the means of local community members (for other examples, see Monk and Ni Luanaigh 2006; Home 2009; Wightman 2010). Another element of the 'incomer impact' is the significant proportion of second and holiday homes which also represent housing removed from local community use. These trends lead to tensions between locals, who are priced out of the housing market in their home area, and the incoming population.

The rate and scale of housing developments on the edge of villages are of concern on the estates and in the estate communities studied. Many interviewees feel that village growth tends to be out of proportion with the existing village size, resulting in a change in village character, increasing commuting, and outweighing perceived local need for additional housing. Others are unhappy with plans for a large volume of new affordable housing, as they believe that this has the potential to import social problems into the rural community. These negative perceptions highlight a further divide between 'incomers' and 'locals'; even where incomers are not stereotypical wealthy second-home owners, locals are still critical of their perceived lack of contribution to the community and local economy, and express concern that current services will not cope with an influx of population. Nonetheless, some housing developments are viewed positively, especially when matched with new employment opportunities where a proportion of planned affordable units will meet housing need and in areas which have lacked development in previous decades.

In general, new residents are considered vital to maintain a village population, as well as to provide critical mass for services and facilities. Community members, however, call for greater prioritisation of 'locals' for affordable housing, in addition to a shift from a reliance on the provision of housing by the estate to a wider range of housing sources through, for example, housing associations or plots for self-building. Further 'knock-on' benefits of rural housing development described by community members include planning gain contributing to village hall restoration, added trade for village shops, and an increased labour pool for the estate.

Employment

Land-based employment opportunities in rural Scotland have steadily declined in recent decades (Ward 2006). This decline emerged as a key theme within the research, with change driven by market forces operating from local to global scales. Locally, community members report a lack of employment possibilities (or 'under-employment') because of the closure or downturn of traditional industries, such as forestry, mining or estate employment, and the changing rural workforce. Several challenges face businesses and self-employed community members in upland areas (Box 5.4).

Box 5.4 Challenges facing businesses and self-employed members of communities in upland Scotland

- Competition from large supermarkets
- Uncertain and costly regulatory/legislative environment
- Lack of funding to encourage local employment
- Lack of childcare
- Lack of land availability to allow development or expansion of industry
- Distance to markets to sell produce, or import, e.g. building materials or animal feed
- Seasonality of employment opportunities

Community members also highlight the uncertainty of local employment provided by only a handful of large employers, and the negative impact were one of these to relocate or close. On the other hand, large employment sources are regarded as essential for maintaining a viable population and key services on some estates. On one estate, a large fish farm is considered a key element in community sustainability: 'if it wasn't for the fish farm we would have no school and bugger all here', asserts a community member. Nonetheless, external and global-scale forces also drive and influence employment on the local scale, having an impact on the resilience and potential sustainability of the local community. Community members recognise that, when controlled by large external companies, the resources of their remote region are not used to provide maximum income locally. Other community members highlight significant environmental damage caused by such large-scale industries. Some businesses, however, have experienced increasing trade because of large-scale trends, such as in tourism. It is therefore likely that communities with a diversity of business types and sizes have greater resilience to external forces.

The case studies also demonstrate the wider social and environmental implications of a lack of local employment, such as an increase in carbon emissions because of commuting. As employment for many in these rural communities requires commuting significant distances to larger settlements, living in the community is considered 'a lifestyle choice', with both positive and negative effects on the community and individuals. In general, the benefits of a growing population were felt to outweigh the negative effects of increased commuting.

On some estates, the number of available jobs is relatively high, resulting in shortages of local labour for employers. This relates particularly to seasonal and low-paid service jobs; despite the abundance of such opportunities, the outcomes are increased employment of foreign immigrants, on the one hand, and community tensions regarding unemployed community members on the other. In other communities, jobs are limited, and community members have few employment opportunities to choose from, with little potential for career development. Many people in rural communities must practise 'pluri-active' employment because of seasonality and low wage levels. Low wages, high house prices, and a perceived lack of opportunities for career development force many younger people to leave their home area. This phenomenon is causing population decreases in these rural areas, with associated impacts on service provision, further dislodging the roots of communities. Thus, housing and employment challenges for young people are a central issue in the minds of community members of all age groups.

The potential for community land acquisition in order to generate local employment was a regular topic of discussion. On several estates, community members consider land reform, with the goal of creating new crofts, as the route to positive progress and a chance to provide the local community with the opportunity to build homes and businesses. Themes of 'limited space' are exemplified, with new residents expressing frustration at being unable to purchase a croft and expand potential land-based income generation, such as growing vegetables for local sale. This desire for croft ownership is disputed, however, by other community crofting members. This difference reflects divergent viewpoints between the local, long-established crofting community and the incoming population regarding the value of ownership, and aspirations in terms of employment and income.

While community members note that crofting does not generate a full-time income, they also perceive a sense of empowerment with a rise in crofting through providing the space for entrepreneurial development and reconnecting rural communities to the land. Many agree that the provision of land, in the style of crofts, would be significant in sustaining the local community. There are also more sceptical views regarding the potential for crofting development, however, especially given its complex legislative framework, lack of income potential, and changing societal aspirations. One landowner feels that a central concern is the individual right to buy which, s/he believes, will break down the crofting structure and result in the end of a 'way of life'.

On one estate, community members are calling for previously de-crofted lands to be used for community-led housing and business development, generating tensions between the landowning family and that section of the community. In such a context,

TABLE 5.2 Characteristics, examples of, and key factors for 'community spirit'

Characteristics of 'community spirit'	Examples demonstrating community spirit	Key factors required for community spirit
Neighbourliness	'Healthy' community groups	Volunteering, including involvement of proactive individuals
Environmental consciousness	Community events and exhibitions	Funding opportunities
Community awareness	Support for children's events and groups	Local confidence and interest
Community activity	Management of community assets	Community achievements and capacity building

the development of co-management structures could perhaps encourage landowners to expand the land available for crofting, facilitating associated employment and community benefits, a point discussed further below.

Community spirit and cohesion

Table 5.2 illustrates the views of interviewees regarding characteristics of, and factors required for, the central – though intrinsically indefinable (see Gilchrist 2000) – concept of 'community spirit'. Related to many of these factors is the success of community groups which have acquired or managed assets, providing a focus for community energy and development potential.

The role of specific individuals is a significant element in community group activity and the generation of 'community spirit'; as one estate manager surmises: 'some people just make community happen'. While community members identify the links between community group activity, community spirit and key initiators within the community – and the potential burdens that can result – they also highlight the negative influence that individuals may exercise. As one estate manager explains: 'I think the trouble is [that] personalities become disproportionate within small communities.'

A decline in community groups and activities reduces interactions among the community as a whole and weakens positive networks. Other threats to, and opportunities for, community spirit are presented in Table 5.3. On one estate, older female community members commented that their local branch of the Women's Rural Institute (WRI) has been replaced, in their view, with television soap operas, which isolate women in their homes. This example is linked to external forces and wider societal change acting on rural communities. Other examples include perceived changes in the expectations and priorities of community members, an increase in single-person households, and an ageing population.

A final issue resulting in debate and division within the estate communities studied is the rise of renewable energy developments, renowned in the literature for generating strong opposition by so-called Nimbys ('Not in my back yard': Warren and McFadyen 2010). Evidence from the estates and estate communities suggests that the

TABLE 5.3 Threats to, and opportunities for, community spirit

Threats	Opportunities
Volunteer fatigue and dependency on key actors	Local newsletters and papers, contributing to community identity
Cyclical community group vibrancy	Activities that focus on children and bringing people together, as well as local history and heritage
Negative community 'attitude' and apathy	Partnerships between community groups
Tensions regarding the representation of interests on community groups (e.g. incomer/retired dominance)	Representation of all community interests within key groups, e.g. community council

opposition tends to emerge from new residents; such people, however, are also often active initiators of community activity and development (including renewable energy opportunities). This challenges simple discourses of preservation versus progress: it is not just the 'incomers' who sometimes advocate little change and want to preserve the environment and property prices; 'locals' do not want change to traditional systems which are perceived to work well. All community members share a sense of enjoyment and place attachment, however, which may be interpreted as conducive to positive community spirit.

On each estate, diverse examples of community tensions and splits were evident: for example, between estate and village communities, crofters and non-crofters, or those for and against development proposals. All such divides may be interpreted as acting against community cohesion. They may also be classified as issues of power – where traditional roles of power and influence are challenged by groups with non-traditional knowledge and skills, exacerbating feelings of powerlessness by long-established 'locals' in changing rural communities. The concept of 'community spirit' may thus be considered complex and dynamic, dependent on the sense of empowerment and on the personal values held by individual community members.

THE ROLE OF THE LANDOWNER IN ADDRESSING COMMUNITY ISSUES

Housing provision and development

The perceived responsibility of the landowner in providing adequate and affordable housing is highlighted through criticisms of a lack of such provision in recent estate-led development. Community members do, however, recognise the financial investment that the estate must make to provide affordable housing developments which may be economically unviable. Estate managers and owners described these challenges in detail, highlighting several key barriers to landowners developing affordable housing. Such developments are often too large a capital investment for the estate business, especially considering the significant cost of building infrastructure in rural areas. The payback period for such an investment is not feasible for some landowners, especially in conjunction with the tax burden on property development. This was a major point of contention among estate owners and managers: 'it's not the landowner that should be at fault if there is no affordable housing because we get taxed any-which-way'. Another landowner describes affordable housing development as 'economic suicide'. A further example of the debates surrounding housing development on private estates is illustrated in Box 5.5.

Box 5.5 A reluctance to release land

Negative interactions relating to housing may occur between estate owners and communities when there has been a lack of action throughout the period of ownership of a particular family or due to issues of community resistance. On one estate, the community's calls for the estate to release land for housing and other developments were a central concern: 'We have been here for 25 years and we've watched . . . other communities grow but . . . [here] the estate has stifled every possibility . . .' (Community member).

The researcher was invited to the inaugural meeting of the 'community land reform group' which aimed to use the lever of the LRSA to 'bring the estate to the table' and force it to provide land for housing, business units and crofting. While the landowner is keen to promote local development, s/he explains that the estate's ownership and decision-making mechanisms prohibit the simple distribution of its assets for free or below market rent:

> . . . we are regularly approached by people who want to get access to land . . . which I'm keen to do as a local resident who wants to see community development. But . . . we have to satisfy the directors that in agreeing to [these land sales] that they're not in contravention of their obligations under the Companies Act. And that's a difficult balance sometimes to strike (Landowner).

At the meeting, members of the community land-reform group dismissed this explanation as a smokescreen, implying that the landowner was reluctant to release land for reasons of primarily financial self-interest. The estate management, however, simultaneously submitted planning applications for a development with a mix of housing types, arguably addressing the call for affordable housing and population increase.

Estate owners convey resentment towards what they describe as the 'layers' of planning requirements and obstructions to obtain permission for new developments, as well as the perceived slow process. The greater the number of layers, the more likely that the requirements and governance structures will be in conflict, explains a landowner grappling with the planning departments of both Highland Council and the Cairngorms National Park Authority. Another estate manager explains that the length of time taken to carry through a large housing scheme, to include a high proportion of affordable housing, has resulted in the loss of support from the housing association, as well as causing community anger to be directed at the estate. Private landowners are also reluctant to pursue affordable housing developments because of the requirements of local authorities for those listed as homeless in the region to be prioritised for available housing, and for landlords to provide long leases. Despite these challenges, landowners highlight potential solutions, including appropriate government support (such as grant provision and taxation relief), and a change in perceptions of the private rented sector among politicians. The estates also show innovative and novel examples of how they could provide affordable housing: for example, provisions secured through the planning system, including Rural Housing Burdens[1] and shared equity schemes.[2] In some cases, community concern centres on the perception that estate-driven development is making a profit for the landowner, to the detriment of the local community (for example, through loss of village character or pressure on services). Negative perceptions of landowners' motivations may ultimately influence community acceptance of developments. Community members also indicate a sense of uncertainty regarding the plans of estates for property sales, development and renting. This lack of knowledge or understanding of landowners' plans or ambitions, among community members and estate management, suggests that landowners may have a 'closed' approach and be reluctant to involve other parties in planning and decision-making. A landowner may also be unsure of her/his own future development plans because of current or potential planning requirements; this may also be linked to a lack of strategic estate business planning, however.

In some cases, estate tourism enterprises remove housing from potential long-term community use. Community members on one estate express displeasure at the sale of houses within a new estate development for use as holiday homes. Private landowners are therefore faced with the dilemma of whether to forgo potential income or capital realisation to retain housing that local people can afford. Holiday housing is a potential and growing income source for privately owned estates, providing capital taxation relief, which is not possible for housing developed for sale or rent; and, to a certain extent, estate-owned holiday homes generate indirect estate employment.

Private estates also have an historic and, in some cases, continuing role to play in providing accommodation for estate workers. The traditional arrangement of tied housing, where employment on an estate also guarantees rent-free accommodation, has been removed as a privilege of employment on several estates; many estate workers are also residential tenants. Both estate management and community members perceive this to be 'modernisation' by the estate. While estates with a policy of retaining housing for ex-employees at nominal rents are seen to demonstrate a sense of social responsibility, with housing shortages these 'pension schemes' are also declining: for example, when available housing is refocused for tourism. While the residents of tied housing would not deny the benefits – to some, a key reason to undertake estate work – other disadvantages (for example, uncertainty surrounding housing security on termination of employment; a perceived 'missed investment' in home ownership; a mismatch between low estate wages and high rural house prices) make this an uncertain form of affordable rural housing.

One estate manager explains that the estate's policy of housing former employees, with 'pension perks' including free electricity and house maintenance, is considered a route to overcoming the problem of low wages preventing estate employees finding secure housing, as well as retaining a workforce on the estate. A further frequent disadvantage of both tied and estate rental properties is their poor state of repair, however, which has a negative impact on affordability because of high heating costs, for example. One community member explains that her rented property is technically derelict; another recounts that the landowner and estate management express their dismay and embarrassment that estate properties are damp and require major roof repairs. Communication regarding maintenance is a key interaction between the landowner and the community; where maintenance plans and priorities are not conveyed effectively to the community, this can cause conflict. Nonetheless, there is a widely held belief among community members that estates still play an important role in providing rented housing and, despite issues relating to housing maintenance, these properties remain key assets for estate and community sustainability.

Developing employment opportunities

Many participants describe both positive and negative implications of the provision of jobs by the estate. Estate jobs often involve multiple roles, with employees required to pick up a variety of tasks. While estates promote the employment of local people, many jobs are low paid, especially for women. The issue of low wages is stressed by community members but they also state that the other 'benefits' of estate work – for example, housing and rural lifestyle – are likely to compensate for low wages.

Employment by the estate is also considered desirable because of perceptions that the estate management often takes responsibility for creating work beyond that of other employers. One local contractor describes the assistance of the estate in providing dependable work to his business. Another estate is criticised, however, for employing external contractors. Such varying perspectives on estate employment reveal a complex picture, perhaps highlighting a lack of communication about estate management planning to the community, but also suggesting that estate employment that maximises local opportunities has a key role in sustaining rural communities. Equally, estate managers and landowning family members believe that estate employment and employee skills strengthen the sustainability of estate businesses.

Many of the estates have a policy of keeping business enterprises and employment potential 'in house', or within the main estate business, rather than relying on contractors or external agencies, mainly to minimise expenditure. Such a policy has both positive and negative impacts in terms of employment: it promotes estate employment, in order to meet estate requirements, yet reduces indirect employment and tenancies that could contribute positively to community sustainability. Local community members often regard employment generation as part of the landowner's role; they also recognise, however, the need for community empowerment in income and employment generation. Box 5.6 provides a useful example of a shift from estate-based employment to self-employment.

Box 5.6 From redundancy to self-employment

One estate presents a positive example of facilitating and providing local self-employment while meeting its objectives. In this case, saddled with significant inheritance tax, the previous landowner was forced to reduce costs in order to retain family ownership: 'He died young . . . and sort of crippled the estate with inheritance . . . His son took the reins on . . . We [estate employees] were a burden to them really, you know . . .' (Community member). Many subsequent redundancies of estate workers were converted to self-employment, however, because the estate facilitated start-up businesses by offering former employees the continued use of workshops and vehicles. The landowner also provided guaranteed work for these newly self-employed tradesmen on the estate; the major change was that the estate could not afford their full-time employment. This transition is described by former estate employees as 'a good opportunity': '. . . you got complete independence . . . It's been hard work, but it's paid dividends. I've probably . . . a more comfortable lifestyle than . . . [if I'd] been still working on the [estate] . . .' (Community member).

The deal ended tied housing arrangements because the former employees who stayed on were required to rent the houses that their employment by the estate had previously provided. This may be perceived as a positive by-product of this change, empowering the ex-employees in a formalised housing arrangement. It could also have resulted in their out-migration, however, if rents had been unaffordable. Several successful businessmen and families continue to live on the estate, in one case in a self-built house.

There are also perceptions of 'over-employment' on some estates, which is not considered productive by community members and landowning family members, and may be connected to tensions between the village and estate community. On the one hand, these estates have the capacity to provide land or premises for local business development and therefore to increase indirect employment. While these estates largely illustrate an open and encouraging attitude to the development of businesses employing local people on estate land, the power of landowners to inhibit and discourage business development is evident on other estates. This limitation of indirect employment is attributed to a lack of available space, as well as to landowners' beliefs concerning the potential markets for small businesses and their impact on the community. Similarly, landowners' decisions regarding the development of crofts have implications for local employment and income opportunities. These attitudes ultimately reflect the landowners' personalities, as well as their motivations; it is therefore likely that such decisions depend on the specific owner rather than on common factors such as regulatory or financial influences.

Overall, there has been a decline in estate employment. Landowners and community members describe times when the estate was a central rural employer, and jobs were numerous. The decline is believed to have been due primarily to business

---

Box 5.7 Encouraging a diversity of estate enterprises through tenancies

On one estate, extensive diversification and the promotion of tenancies may be considered positive contributions to rural community sustainability. The numerous small businesses present on the estate, and the estate's role in them are described:

'. . . what we've tended to do is we've now hived off things – we've started . . . the granite quarry . . . [and] a fish hatchery . . . they're a tenancy – but again we work pretty closely together' (Landowner).
'. . . it's got to work . . . if you take into consideration the strawberry farm, the riding stables, the quarry, the fish farm . . . [around] 50 people rely on a job on the estate' (Estate management).

This estate management system has been positive for local employment, empowering the community through support for small businesses, and providing income sources for community members and for the estate. The encouragement of tenancies is considered primarily a form of diversification, in addition to risk spreading and maximising land-use productivity for the estate. The motivating factors for encouraging tenancies are not driven entirely by responsibility for the community, however, but rather by the need to ensure estate financial viability: '. . . we are always open to business ventures . . . [but] it has to fit in with the running and management of the estate' (Estate management). Nonetheless, members of the estate community recognise the wider socioeconomic benefits: '[Landowner] has actually stuck his neck out . . . He has allowed people to come in and do things . . . he's given people opportunities . . .' (Community member).

The landowner views the development of estate community enterprises and tenancies as fundamental to estate sustainability, as financial viability is vital for continuing family ownership. A reportedly 'open' attitude, however, probably encourages two-way interactions between the community and the estate regarding employment opportunities and supporting start-up businesses. This landowner exhibits a pragmatic attitude to such interactions, rather than a sentimental or 'paternalistic' approach.

---

efficiency and financial requirements, combined with the long-standing trend of reducing numbers of land-based workers through modernisation and mechanisation. Others, however, reiterate a perception of community dependency on estate employment and a lack of confidence in self-employment. Community members also perceive that a decline in direct estate employment has been matched with an increase in tenancies, including former farming employees becoming agricultural tenants of the estate. These examples reinforce the view that increasing estate tenancies (including agricultural and commercial) is potentially a more secure and productive method for the estate to maximise business output, rather than maintaining operations 'in-house' and with responsibility for a workforce. Box 5.7 provides an illustration

---

Box 5.8 Does a focus of estate management on field sports imply a lack of jobs?

Motivations for estate ownership, land-use decision-making and the personality of the landowner are significant in the provision of direct and indirect estate employment. On one estate, because of the owner's keen interest in sport shooting, activities and employment focus on game management. Farmland is tenanted, forestry work is undertaken by contractors, and a local firm services the estate-owned holiday cottages. Furthermore, there is no local estate office: an external agency provides land management services. Therefore, while the estate undoubtedly supports significant indirect estate employment, it directly employs only two gamekeepers locally. As one community member asserts:

> 'They need to change the attitude, I suppose [Landowner] looks at this as it is – running . . . a sporting estate and let's keep it that way because it doesn't need much effort on his part, whereas if someone said "why aren't you using this asset to run it as a business?" . . . he would have to spend a lot more time and effort, money and a lot of hard work.'

The reason for this narrow focus of employment, in addition to personal interest, may also be that this landowner does not live on the estate. Despite a close working relationship with many community members, a lack of day-to-day local experience may inhibit a landowner's understanding of development and employment opportunities. An absentee landowner may also rely on 'gatekeepers' for community information and ideas, such as estate employees, who may pass on subjective and narrow perspectives. Nonetheless, the local benefits and indirect employment generated by game management and sport shooting were welcomed in this community.

---

of estate diversification through tenancy generation, emphasising a changing 'estate community', as tenants are unlikely to have as close a connection to the estate as direct employees. It may therefore be inferred that changes to estate employment structures, with the goal of business efficiency, have contributed to the creation of new relationships or levels of influence between landowners and their 'tenant' estate communities.

Further examples illustrate the potential for growth in direct and indirect estate employment (Box 5.8 provides an illustration). Employment expansion may be enabled by an estate providing facilities such as business premises but this ultimately depends on the landowner's motivations, decisions, level of involvement in day-to-day management, and vision for the estate. Nonetheless, both estate management and community members recognise that fulfilling the potential and increasing employment opportunities are challenges owing to several key factors and uncertainties, including the existence of a 'rooted' community with strong ties to the local area from which to obtain estate employees, and insufficient estate income. A common theme regarding the potential to reverse declining rural employment is the extent to which private landowners interact with the estate and village community, and the processes of engagement used, as discussed below.

PLAYERS, PROCESSES AND POLICY: THE POTENTIAL FOR PRIVATE LANDOWNER AND ESTATE COMMUNITY PARTNERSHIPS

This research illustrates a great diversity of partnerships between estates and communities, and the potential for further partnership working (see Box 5.9). In particular, there is widespread evidence of a breakdown of traditional hierarchies because of societal change and estate modernisation, increasing the accessibility of landowners and the extent of interactions between the estate and community: for example, through the formalising of tenancies or by providing formal employment contracts. Such examples may be considered illustrative of community empowerment and a shift in power structures away from historic paternalism. While this changing status is welcomed by some landowners, it is not by others, highlighting the fact that some perceive the current legislative framework as a threat to private landownership. This final section considers the key factors and opportunities for partnership working, and the practical and institutional constraints inhibiting progress.

Box 5.9 Defining 'partnerships' in the estate community context

Typically, rural partnerships involve representatives of public, private, voluntary and community interests who share a degree of commitment to specific objectives, at a strategic or delivery level (Derkzen and Bock 2009). On the estates studied, community members and estate management define 'partnerships' as including informal arrangements through which estates provide community amenities and services, as well as the co-management of assets by estates and community groups. These partnership opportunities include property rental, the provision of estate property or facilities for community use (through either donation or rent), the provision of access and path networks, the active involvement of the estate with community groups, farming partnerships (including machinery rings), employment, and other formal arrangements, such as crofting tenure. Formal business agreements and tenancies are also recognised as partnerships.

Breaking down traditional hierarchies

An initial opportunity for partnership development between private landowners and communities involves the perception of a less marked traditional landowner–estate community hierarchy, through estate and wider societal modernisation. Landowners and community members both believe that modernisation includes, crucially, a change in the social distance between landowner and community. For example, one community member describes the change since her childhood: then, you were never to enter the front door of the landowner's home; today, she is happy to 'go in and shout for [him/her]'. Landowners are described as a positive institution with which the community wishes to retain links, and relationships are improving as involvement increases: 'Now it is a completely relaxed relationship . . . the attitude has changed' (Community member).

Other aspects of modernisation include a changing rural population with significant in-migration (for example, of retirees and commuters), the abolition of feudalism, a breaking down of 'laird' stereotypes, and arguably a reduction of landowners' historical power: 'A hundred years ago, the laird's word was law . . . Now they virtually have no say' (Community member). Landowner participants consider this changing status and associated power relations, on the one hand, a progressive step: '. . . one of the really good things that's changed is . . . a huge shift away from the sort of paternalistic involvement . . . between estates and the community to a much more realistic, pragmatic relationship' (Landowner).

On the other hand, the changing relationship is considered compulsory yet potentially troublesome: '. . . you can't work on feudal systems these days . . . with legislation the way it is . . . we are in a period of transition . . . it is not straightforward' (Estate management).

Positive communication and the 'face' of the estate

Communication is key to developing partnership opportunities between the estate and the community. Communication practices that are perceived to contribute positively to partnerships between estates and communities include estate newsletters, complaints and praise forums, estate surgeries, and 'long-service' awards. A two-way process of understanding and listening is critical, including appropriate, genuine and ongoing consultation and an 'open-door' policy: a fundamental approach to ensuring continued dialogue and for breaking down historical barriers between landowner and community. Communication by the community to the landowner is also central to the partnership approach, highlighting that it is a shared responsibility.

Community members express discontent, however, when they consider the estate to be interfering: for example: '[there's] ill feeling that they've always got a finger in every pie' (Community member). It may be inferred that the preference of community members is for more distant interaction in combination with an 'open door' and friendly attitude when problems arise: for example, when a tenant's roof needs maintenance, '. . . you just go down to the office and . . . say I'm needing another slate . . . I mean they don't bother about you otherwise, you just get on with life . . .' (Community member).

The 'face of the estate' is crucially important in building positive relations and community engagement. This is ultimately related to the accessibility and the personality of the landowner and estate managers. A key personality trait in this regard is 'approachability'; examples include landowners who are referred to by their first names by community members, are integrated socially with the community, and respected for being 'accessible'. Such characteristics may be considered as indicating increasing equality and the breakdown of historical class differences between landowner and community. Effective processes therefore depend on individuals, their backgrounds, values and, at times, their confidence. Community perceptions identify whether the landowner is perceived as a good local 'ambassador' or 'patron', however, or is considered to possess a 'fair' attitude. For landowners and for community members, pragmatism and realism were considered personal characteristics conducive to partnerships.

Where estate representatives, including the landowner and landowning family, have the types of characteristics described above, this appears conducive to effective engagement or partnership working but the opposite can also occur:

'We did start a programme of estate meetings – interactions with the village . . . a formal one, which I haven't continued . . . Quite simply . . . the chairman of the company was deployed to chair those meetings and he's not somebody I think engages well in that context . . . He's a toff with a very posh accent . . . It doesn't convey the kind of values that I think the estate needs to convey'. (Landowner)

In addition, some estate managers believe that a 'healthy separation' is required for them to undertake their professional role, implying that partnerships require a pragmatic and professional relationship between estate and community.

Proactive involvement

The research did not show that resident and absentee landowners differed in approach or opportunity for partnership working. This may be due to the focus of the research on estates that, in their response to the initial landowners' survey, emphasised positive community involvement. The proactive attitude to partnership approaches demonstrated by both resident and absentee landowners, however, emphasises the role of individual personality over estate type or ownership situation. Absentee case-study owners are described:

'No, s/he is not here a lot but s/he is remarkably well informed about the area . . .'. (Community member)
'. . . from a home base not even within reasonable travelling distance of the community . . . [S/he] keeps [his/her] fingers on the pulse of the estate, by maintaining contact with groups of people that are important in the community'. (Community member)

This suggests that access to the landowner and estate management through proactive communication may be considered more significant than whether the landowner is resident or not.

The estate manager (or 'factor') has a similarly challenging, but crucial, role in facilitating positive interactions and potential partnerships. The estate managers in this study have considerable knowledge and understanding of the local community because most are members of the community themselves, many with active involvement in its social life. They occupy an awkward position, however, with a personal stake both in the estate business and in community sustainability. Ultimately, the landowner sets the tone for community engagement but, undoubtedly, other estate management personalities have an impact, positive or negative, on these processes. For example, one estate manager suggests that broad community support for estate activities may be due to his own frequent direct interaction with community members because he lives in the area, while another community member believes that: '. . . the bottom line is that [he] is here, and that if people want to speak to [him] – go and find him . . . [he is] a very approachable guy'.

The involvement of land management agencies is also contentious in terms of partnerships and promoting sustainability. While some perceive such agencies as a positive body of expertise with broader perspectives on contemporary land management, others: describe agents as showing no interest in community issues; question their loyalty to the estate, community or commercial interests; and describe barriers created by their status as 'non-local', reinforcing local–incomer divides. In such cases, community members tend to approach directly the landowner or other resident estate employees with concerns, bypassing the external estate manager. The use of an external estate manager to 'tell bad news' results in a loss of respect for the landowner. Again, however, the personality of individual estate managers and their style of interaction with community members may be more significant than whether they are employed by the estate or by an external agency.

Resource and skill limitations

The processes of establishing and maintaining partnerships between estates and communities are sometimes constrained and face certain challenges. Partnership development and engagement between the estate and the community require significant effort and time commitments: for example, in attending community group meetings. Partnerships are understandably limited where the landowner and/or estate management lack time or the practical and personal ability to lead or undertake community engagement processes, or do not have the necessary financial or other essential resources for such processes.

A change in ownership may also be a crucial factor. For example, on estates where community members considered previous landowners and estate management as more charismatic than current ones, this appears to have a negative impact on the current landowner's relationship with the community: '. . . [he] is quite solitary and quiet, and a thinker . . . it took a long time for people to stop comparing him to [previous landowner]' (Estate management). In this example, community members perceive the quiet trait of the landowner as a lack of interest in promoting community interests, though this is counter to this landowner's self-expressed motivations. Another landowner is a self-confessed 'recluse'. Such social distance and resulting negative perceptions are likely to be constraints to effective partnership development between estate and community.

Disconnection, apathy and uncertainty

Partnership working is also constrained where those responsible for managing the estate – including the landowner, employees, and remotely located estate company directors or trustees – are not familiar with, or are considered distant from, the local community. Community members express discontent when they feel distant from management decisions and believe that such decision-making by remote govern-ance lacks innovation, local knowledge, risk-taking approaches and accountability. At times, such external influence threatens to be inconsistent with the objectives expressed by 'local' estate management:

> '. . . even though the estate has had a formal policy agreed at the board level that we should be encouraging land sales to encourage repopulation . . . if you go and talk to people they will say historically [the estate] will not sell land . . . Somewhere between the board and the community there's been a complete breakdown of communications . . . That's got to be resolved'. (Landowner)

This emphasises the importance of communication between all levels of estate man-agement and the need for effective community engagement in estate management goal-setting and planning. Improved communication, in particular face-to-face dialogue, is a key factor in improving relationships.

In one case, the separation of a village from its neighbouring estate because of modernisation and development led to community members stating that they have little interaction with estate management and estate employees. Community members resident in the village explain that their perspectives on the influence of the estate are different from those of 'estate community' members (see Box 5.10 for an example of different perspectives). Furthermore, where there is little perceived interaction by estates with communities, because of a small number of tenancies, for example, some community members may not believe any relationship exists or 'feel the estate and community run parallel'. Community members on a number of estates believe that other institutions and external forces – including local authorities, national govern-ment policy and wider societal norms – have greater influence than the estate and landowner.

Community apathy – and related issues of volunteer fatigue and community cohesion – can be a major constraint to developing and maintaining partnerships. As an estate employee states: 'You know there's all this talk about "oh well, you know, they don't help us", but actually, when help is offered . . . people don't necessarily take it.' Similarly, estate management may express a sense that the local community does not support the estate, in particular, new residents who lack awareness of the estate's activities.

A related issue is the level of expectation, within a community, of involvement and interaction with the estate and its management. On many estates, community members strongly assert that the estate community primarily expects engagement only where estate management plans will affect them: 'I feel that they only need to share the decisions that affect us'; 'I don't feel it is my business'. One interpretation

> **Box 5.10** New voices are 'louder'/the power of the crofters
>
> When community members feel that the landowner prioritises communication and support for only one 'fraction' of the community, this can negatively influence relationships between the estate and the community. On one estate, the landowner maintains a close working relationship with the crofting community, and is clearly supportive of non-traditional crofting activities that generate indirect employment. Those outwith the crofting community, however, do not report such an open attitude to business ideas by this landowner: 'Without being unkind, we don't find [Landowner] very helpful' (Community member). Non-crofters assert that the crofters have a 'very big say' and are 'really in a very privileged position' (Community member). Conversely, more articulate new residents are more likely to attend estate-community consultation events on a different estate: 'Aye, you will usually find that them that's come into the village are more free to talk . . . The traditional [village] people . . . the farmers and that – they don't usually attend these things' (Community member). These examples highlight the tensions and power struggles between local 'traditional' residents and newer arrivals, and the potential for landowner and estate-community involvement to inadvertently reinforce these divides through apparently preferential engagement and support. The quotations above also demonstrate the key role of the landowner in ensuring that engagement between the estate and the community is unbiased and with all members of the community. Similarly, it seems important for all potential, viable business ideas, which boost local employment and income sources, to be recognised and supported – whether generated by a 'local' or an 'incomer'.

of such views is that they reflect community apathy flowing from a sense of disconnection between estate and community, therefore impeding potential partnerships. Nevertheless, understanding these views may assist estate management in planning community engagement processes and initiating partnerships where interest is expressed and mutual benefit may be recognised.

Constraints also appear where members of a community express uncertainty about estate-management plans and future landownership. This appears to have a limiting and unsettling impact on those involved in current and potential tenancies, and in future partnerships: 'I fear what happens when [Landowner] dies . . . I don't know to what degree . . . [successor will] have the same philosophy and the same views . . . And will [s/he] want businesses to be structured the way they are now . . . ?' (Community member).

The landowners involved also express their own uncertainties, including the possible impacts of shifts in public perceptions, politics, external funding sources and land-reform legislation. Such uncertainties are having a detrimental impact on the potential for creating new agricultural tenancies and new crofts on estate land, with landowners concerned that legislative reform will lead to stronger right-to-buy provisions. Conversely, at the local level, negative perceptions of private landownership persist, including inaction and indecision, absenteeism, the persistence of feudal attitudes, and negative paternalistic relationships.

Factors contributing to the persistence of these perceptions include a lack of 'honest answers' in community engagement and of awareness by the landowner of community views. A recurrent view was that landowners avoid genuine community engagement because of an apparent need to retain 'control'. This may result from individual personality traits, or may be indicative of a wider reaction by private landowners to external threats to private property rights, such as land reform and the Scottish Government's community-empowerment agenda.

Inequality and power relations

Fundamentally, this research demonstrates that the key constraints to partnerships between the landowner, estate and community relate to inequalities in power relations. The powers of private landownership include local economic development, with social and environmental implications, as discussed throughout this chapter and in Chapter 3. A lack of proactive estate-community engagement and communication can also disempower communities. Estate benevolence can have similar effects. In many cases, estates make positive contributions to community activities: examples include landowners' provision of land and financial assistance to community activities and individuals, including large gifts of property for community use. Both community and estate representatives state, however, that the reliance of the community on landowner benevolence may be counterproductive to partnership working, describing the negative corollaries of community expectations and estate benevolence: for example, 'I don't want communities to see the private landowners as money pots . . . It should be about involvement . . . It's not just, they are there, and you have nothing to do with them and you approach them when you want money . . .' (Community member). Indeed, it may be argued that such relationships maintain an unequal paternalistic relationship between landowners and estate communities where landowners' views are seen as superior to community perspectives owing to the power demonstrated through benevolence.

The decisions made by landowners and estate management regarding financial or other community support are, of course, not always motivated by altruism. Estates may engage with, and provide support to, communities in order to protect their business interests, and community members express a sense that such community engagement is 'stage managed' to meet best the needs of the estate, avoiding difficult questions: for example, regarding strategic estate planning and future landownership. They explain that, while their local landowner is willing to provide support, s/he also has their 'own ideas' but that: 'you can change [Landowner] sometimes if you put enough charm into [Landowner]' (Community member). This dependence on landowners' will and personality represents another weak link in 'sustainable' partnership working, with 'charm' being arguably a further precarious element.

Despite the removal of feudalism in the Abolition of Feudal Tenure Etc. (Scotland) Act 2000, community participants continually discuss engagement and potential partnerships as if it were still in place. A sense of feudalism is closely linked to a perception that the 'traditional laird' relationship with the community continues, especially among older generations. Community members also highlight the social difference between the landowner and the wider community, again an indicator of persistent historical roles. This social difference is often not explicit but may be considered an intrinsic element of long-established estate-community interactions: '. . . there is a certain residual kind of loyalty to a Laird idea . . . It's partly work . . . and it's partly deference . . . It's not exactly "tugging the cap" deference but . . . to disagree is not something you'd do' (Estate family member).

This in-built social difference may pose a challenge to balancing power in partner-ships between community members on the one hand, and owners and managers of private estates on the other. Such traditional relationships are maintained by older generations, however, and therefore are likely to diminish with time. This implies a potential for increasing equality, with a shift from traditional social structures to relationships that are less formal and hierarchical, with likely positive implications for partnership working. A related change is clearly the rebalancing of power as a consequence of land reform:

> '. . . there's getting to be stronger people within the community that know more about legislation and people's rights – like just this whole land reform thing, you know – there are more and more people now that are actually paying attention to what's going on everywhere else . . .' (Community member)

This quotation also indicates, however, a possible shift away from working with the estate and, instead, asserting community and individual rights so that the mutual benefits of partnership working may be overlooked or undermined. Crucially, despite the many examples of positive engagement and the potential for partnership approaches, a fundamental weakness of private landownership is that partnerships between an estate and the local community may be unavoidably unequal owing to the greater economic power of one party as land 'owners'. The LRSA gives communities power to redress this perceived power imbalance. Therefore, private landowners who wish to retain asset ownership and minimise community motivations for an estate 'buyout' would be wise to continue to interact with the community and adopt positive approaches to engagement and partnership working: 'You shoot yourself in the foot if you forget about that relationship' (Community member).

In conclusion, the examples from these six estates illustrate the existence of, and potential for, partnerships, including informal arrangements for estates to provide community facilities and support, and formal arrangements of employment and tenancy. Such partnership opportunities are encouraged by modernisation, a break-ing down of historical hierarchy, estate-community communication, and a perceived approachability of estate management and the landowner by the community, in addition to a pragmatic – rather than sentimental or deferential – relationship. The development of partnerships may be constrained by resource, time and personality limitations; a sense of estate-community disconnection with modernisation; com-munity apathy; negative perceptions of the landowner; uncertainty; and a perceived persistence of feudalism. A fundamental constraint which emerges is the sense of an imbalance of power in potential partnerships, with an awareness of the implications for private landownership of a shift to greater equality on the part of the community: a shift that highlights the crucial roles of (i) effective estate-community engagement, and (ii) the motivation of the private landowner to ensure positive relationships with the community.

IMPLICATIONS FOR POLICY AND PRACTICE

Given the benefits of partnerships between private estates and local communities, and the widespread desire for them, there is an opportunity for government support to maximise the potential for such partnerships to contribute to rural sustainability. For example, there is a clear role for external assistance in ensuring that the 'nuts and bolts' of partnership working (in addition to engagement processes) are in place, such as the time required for the key actors to identify and develop partnership opportunities. This may be supported through external facilitation (for example, where a third party brings together estate and community representatives to discuss future developments) or perhaps through financial resources to 'kick-start' opportunities. Similarly, the development and promotion of specific guidance on engagement processes and partnership mechanisms for landowners, estate managers, and community members would also contribute positively to the delivery of partnership approaches (such guidance can be found in Glass et al. 2012).[3] Reluctant landowners or apathetic community members may be incentivised to develop partnerships through the promotion of 'peer successes', and evidence of positive impacts, for example through the use of sustainability indicator sets (see Chapter 8) and economic assessments. Government may have a key role to play in recognising the positive impacts of partnership approaches which may be incorporated in the current review of land-reform policy in Scotland (2012–14).[4]

New approaches may be developed to assist in estate–community partnership development, as intermediate approaches to governance between the polarised positions of private and community ownership. Indeed, several estates are now owned and managed by partnerships involving local people and national conservation organisations, notably the John Muir Trust (Warren and McKee 2011) (see Chapter 7 for more detail). Support is also required to overcome barriers (including the uncertainty of legislative change and taxation levels) to tenancies and joint enterprises by creating flexible ownership mechanisms that encourage shared management and equity. Such support could contribute positively to the British government's localism agenda, and the development and eventual implementation of the Scottish Government's Community Empowerment and Renewal Bill. In particular, there should be an emphasis on involving the private sector in community-development processes (supported by appropriate financial assistance), and creating institutional capacity to work with 'enabled communities' (see Skerratt 2010). Estate–community partnerships could contribute to these goals, and therefore would be eligible for government assistance. The apparent negative perceptions of private landownership held by the public,

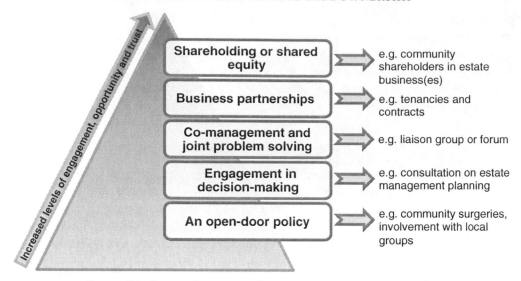

FIGURE 5.3 *Suggested spectrum of community engagement approaches*

the media and certain political classes (for example, as discussed in Chapter 3) remain a key challenge. There may be a reluctance for the government to provide support to private landowners where it is perceived that estate businesses and wealthy landowners are gaining from further taxpayer assistance. This reinforces the need for further investigation into the contributions of partnerships to sustainable rural development, in order to inform public and political perceptions.

Similarly, private landowners may be encouraged to consider the possibility of formalising community engagement processes by developing novel partnerships within the estate business or shareholding schemes (see Figure 5.3). The latter option would involve community members investing a financial stake (either personally or within a representative body) in the estate business, thereby acquiring the right to contribute to decision-making processes, as is standard with any company shareholding scheme. The estate business should benefit financially from shareholder investments, used to develop further income streams and underpin estate economic sustainability. A shift to estate shareholding schemes may lead to an altered taxation environment, however, which may not be as favourable for privately owned estates owing to the introduction of business rates payable on estate companies, for example, or land value taxation (see Wightman 2010). Furthermore, landowners may be reluctant to enter into shareholding schemes that would result in a loss of personal or family control of the estate business. Such an approach would require community capacity and pragmatism, including overcoming prejudices and moving beyond a dependency culture. Finally, though private landowners will inevitably be disempowered to some extent by partnership approaches which empower communities, power sharing may contribute significantly towards the legitimation of private landownership in Scottish society, with great potential benefits for all parties. Best-practice recommendations arising from this research for private landowners and rural community members are presented in Box 5.11.

Box 5.11 Best-practice recommendations

Private landowners
- Support community sustainability through the creation of commercial and residential tenancies, in addition to other partnerships.
- Follow good practice in community engagement processes in order to promote estate and community sustainability, including ensuring transparency, accountability and an ongoing dialogue with estate 'stakeholders'.

Communities
- Promote community cohesion through welcoming incoming skills, developing joint visions and embracing opportunities for community development: for example, through diversifying employment opportunities.
- Professionalise relationship with private estates in order to reduce inequality, estate dependency and perceptions of apathy.

A key overall finding of this research is that private landowners can play a range of roles in facilitating sustainable rural communities. These roles may be characterised as that of 'contributor', 'enabler' and, of course, 'partner'. The role of 'contributor' promotes community sustainability through estate support (both financial and in kind) for community activities and development. The role of 'enabler' can encourage community enterprise development and affordable housing, contributing to community capacity building and reducing the community's sense of dependency on the estate. As discussed throughout this chapter, the third role, of 'partner' to the community, may be established through developing partnership opportunities in terms of asset co-management, joint estate–community ventures, and shared equity schemes. Such novel partnership approaches have been little explored in the academic literature, though they exist in practice (see Warren 2009; Warren and McKee 2011). Each role implies a negotiation of power between the landowner and the community. It is likely that, with increasing public acceptance of partnership approaches, and the potential positive roles that private landowners can play, community empowerment will be promoted. As suggested above, it may be argued that Scottish private landowners may regard a measure of their own disempowerment as a 'price worth paying' as a way to reduce the pressure for further land reform and to counter other threats to private property rights. Their adoption of this role is likely to have considerable benefits for estate community sustainability, especially by increasing community involvement in the key resource of land.

NOTES

1. The Rural Housing Burden mechanism seeks to 'lock in' the affordability of a rural house/ house plot which is sold below market price through the addition of a legal obligation in the property's title deeds that ensures that, whenever it is sold in the future, the original discount can be retained in perpetuity (for further information, see Alexander 2011).

2. 'Shared equity schemes' refer to funding and housing-tenure mechanisms that involve part ownership by the occupier and another body, such as a housing association or private business. On the estates, shared equity schemes were proposed to ensure that estate-built property remained 'affordable' in the long term and that the estate had a pre-emptive right to buy, were the property to be sold by the part-owner-occupiers. The Scottish Government centrally manages and funds two forms of shared equity scheme, supporting individuals to home ownership, and encouraging housing developers (see Berry 2012 for more information).

3. Further discussion of community engagement in private estate management can be found in a short, practical booklet entitled 'Working Together for Sustainable Estate Communities: Exploring the Potential of Collaborative Initiatives between Private Estates and Communities' (Glass et al. 2012). A draft of the booklet was presented at three workshops in upland Scotland in October/November 2011; following feedback on its content, the final version is available from <www.perth.uhi.ac.uk/sustainable-estates>.

4. A Land Reform Review Group is, at the time of writing, conducting a review of the Land Reform (Scotland) Act, for the Scottish Government. Its remit includes identifying the barriers to communities in gaining access to land and its management. For further information, visit: <http://www.scotland.gov.uk/About/Review/land-reform/Remit> (last accessed 8 January 2013).

References

Alexander, D. (2011). 'The Effectiveness of Rural Housing Burdens'. Carnegie UK Trust, Dunfermline, November 2011.

Berry, K. (2012). 'Affordable Housing Supply Funding'. SPICe Briefing, The Scottish Parliament, Edinburgh, 18 July 2012. <http://www.scottish.parliament.uk/parliamentarybusiness/53169.aspx> (last accessed 12 December 2012).

Bird, S. (1982). 'The impact of estate ownership on social development in a Scottish rural community'. *Sociologia Ruralis* 22, pp. 36–48.

Bridger, J. C. and Luloff, A. E. (1999). 'Toward an interactional approach to sustainable community development'. *Journal of Rural Studies* 15, pp. 377–87.

Bryden, J. and Geisler, C. (2007). 'Community-based land reform: Lessons from Scotland'. *Land Use Policy* 24, pp. 24–34.

Buccleuch, Duke of (2005). 'Modern estates are dusting-off their image'. *The Scotsman*, 28 May 2005.

Derkzen, P. and Bock, B. (2009). 'Partnership and role perception, three case studies on the meaning of being a representative in rural partnerships'. *Environment and Planning C* 27, pp. 75–89.

Fawcett, J. and Costley, N. (2010). 'Public Attitudes towards Estates in Scotland: Results of Quantitative Research Findings'. George Street Research, March 2010.

Fraser, E. D. G., Dougill, A. J., Mabee, W. E., Reed, M. and McAlpine, P. (2006). 'Bottom up and top down: Analysis of participatory processes for sustainability indicator identification as a pathway to community empowerment and sustainable environmental management'. *Journal of Environmental Management* 78, pp. 114–27.

Gilchrist, A. (2000). 'Design for Living: The Challenge of Sustainable Communities'. In: Barton, H. (ed.), *Sustainable Communities: The Potential for Eco-Neighbourhoods*. Earthscan, London, pp. 147–59.

Glass, J. H., McKee, A. and Mc Morran, R. (2012). 'Working Together for Sustainable Estate Communities: exploring the potential of collaborative initiatives between privately-owned estates, communities and other partners'. Centre for Mountain Studies, Perth College, University of the Highlands and Islands.

Higgins, P., Wightman, A. and MacMillan, D. (2002). 'Sporting estates and recreational land-use in the Highlands and Islands of Scotland'. Report for the Economic and Social Research Council.

Home, R. (2009). 'Land ownership in the United Kingdom: Trends, preferences and future challenges'. *Land Use Policy* 26 (1), S103–S108.

Kerr, G. (2004). 'The Contribution and Socio-Economic Role of Scottish Estates'. Summary Report, January 2004. Scottish Agricultural College, Edinburgh.

Land Reform Policy Group (1999). 'Recommendations for Action'. Scottish Office, Edinburgh.

McIntosh, A., Wightman, A. and Morgan, D. (1994). 'Reclaiming the Scottish Highlands: clearance, conflict and crofting.' *The Ecologist* 24 (2), pp. 64–70.

Monk, S. and Ni Luanaigh, A. (2006). 'Rural housing affordability and sustainable communities'. In: Midgely, J. (ed.), *A New Rural Agenda*. IPPR North, Newcastle-upon-Tyne, pp. 114–35.

Samuel, A. M. M. (2000). 'Cultural Symbols and Landowners' Power: The Practice of Managing Scotland's Natural Resource'. *Sociology* 34 (4), pp. 691–706.

Shucksmith, M. and Dargan, L. (2006). 'Skye and Lochalsh: Nature Protection and Biodiversity'. EU CORASON project WP5. University of Newcastle-upon-Tyne.

Skerratt, S. (2010). 'How are Scotland's rural communities taking ownership of their own future?' In: Skerratt, S., Hall, C., Lamprinopoulou, C., McCracken, D., Midgley, A., Price, M., Renwick, A., Revoredo, C., Thomson, S., Williams, F. and Wreford, A., *Rural Scotland in Focus 2010*. Rural Policy Centre, Scottish Agricultural College, Edinburgh, pp. 42–51.

Thomson, S. (2012). 'How have Scotland's rural population, economy and environment changed since the 2010 report?' In: Skerratt, S., Atterton, J., Hall, C., McCracken, D., Renwick, A., Revoredo-Giha, C., Steinerowski, A., Thomson, S., Woolvin, M., Farrington, J, and Heesen, F., *Rural Scotland in Focus 2012*. Rural Policy Centre, Scottish Agricultural College, Edinburgh, pp. 9–26.

Ward, N. (2006). 'Rural Development and the Economies of Rural Areas'. In: Midgley, J. (ed.), *A New Rural Agenda*. IPPR North, Newcastle-upon-Tyne, pp. 46–67.

Warren, C. R. (1999). 'Scottish Land Reform: Time to get Lairds A-Leaping'. *ECOS* 20 (1), pp. 1–15.

Warren, C. R. (2009). *Managing Scotland's Environment*. Second Edition. Edinburgh University Press, Edinburgh. 432 pp.

Warren, C. R. and McFadyen, M. (2010). 'Does community ownership affect public attitudes to wind energy? A case study from south west Scotland'. *Land Use Policy* 27 (2), pp. 204–13.

Warren, C. R. and McKee, A. (2011). 'The Scottish Revolution? Evaluating the impacts of post-devolution land reform'. *Scottish Geographical Journal* 127 (1), pp. 17–39.

Wightman, A. (1996). *Who owns Scotland?* Canongate, Edinburgh. 176 pp.

Wightman, A. (1999). *Scotland: land and power – the agenda for land reform*. Luath Press, Edinburgh. 128 pp.

Wightman, A. (2010). *The Poor Had No Lawyers: Who owns Scotland and how they got it.* Birlinn, Edinburgh. 320 pp.

PART THREE

Perspectives from community and NGO landownership

CHAPTER SIX

Community landownership: rediscovering the road to sustainability

Robert Mc Morran and Alister Scott

INTRODUCTION

This chapter reviews the historical context of community landownership in Scotland, with a theoretical exploration of the sustainability credentials of this form of tenure. Four in-depth case studies from north-west Scotland are used to illuminate the impacts, processes and experiences of community landownership in practice, including the identification of barriers and opportunities for progressing towards sustainability. Five key themes emerge from these case studies: (i) narratives of sustainable rural development; (ii) rebuilding community capacity; iii) (redefining participatory governance and partnership working; (iv) building a framework for economic development; and (v) reconfiguring community–natural resource relationships. These are followed by a discussion relating to the wider relevance and policy implications of the research.

THE EMERGENCE OF COMMUNAL LAND TENURE IN SCOTLAND

Scotland has one of the most concentrated patterns of private landownership in the world, with the current dominance of large private estates a legacy of the longevity of feudal tenure (Wightman, 1997). In recent decades, one response to this situation has been increased demand for community ownership of land, driven by issues of insecurity, neglect and disempowerment, set within the wider macroeconomic climate of decline in community populations, investments and services (MacPhail, 2002; MacAskill, 1999). Community 'buyouts' of land gradually became a rallying cry as part of endogenous community-development activities and programmes (Skerratt 2011; Hunter 2012). The first formal community acquisition of land was in 1923, when Lord Leverhulme gifted his Lewis estate to the local community-based Stornoway Trust (Boyd 1999). It was not until the 1990s, however, that significant momentum began to build, with the purchase in 1993 of the 21,300-hectare North Lochinver Estate for £300,000 by the Assynt Crofters' Trust. This proved to be a catalyst for smaller-scale crofting buyouts, in the townships of Borve (1993) and Melness

(1995) (Brennan 2001; Chenevix-Trench and Philip 2001), igniting wider interest around land reform and community ownership of land. This was encapsulated on the Isle of Eigg and in Knoydart, where local dissatisfaction with perceived irresponsible private landownership led local residents to initiate legal proceedings to buy the land (Dressler 2002; Boyd 2003).

Following the Eigg (1997) and Knoydart (1999) buyouts, the process moved from being relatively ad hoc to a more formalised and legitimised position, with the establishment of the Community Land Unit (CLU) within Highlands and Islands Enterprise (HIE) in 1997. This development signalled a shift in government support for community landownership, with the CLU tasked with providing communities with technical advice relating to the purchase and management of land (SQW 2005). The Scottish Land Fund, funded through the Lottery, was also established in 2001, to provide financial support for community groups to purchase, manage and develop land. The culmination of this was the landmark Land Reform (Scotland) Act 2003 (see Chapter 3). Further buyouts followed, including the Isle of Gigha (2001), North Harris (2003) and South Uist (41,000 hectares) in 2006 (Mackenzie 2006, 2012; Satsangi 2009; Warren 2009).

Community buyouts have now become embedded in the governance of Scotland, with environmental NGOs, local government and non-departmental bodies all playing key roles in the process. These partnerships have often been necessary to enable communities to access the necessary funding and expertise to buy land and establish the governance structure required in the legislation (Bryden and Geisler 2007). In many cases, this has resulted in such organisations becoming formal partners within subsequent community land bodies. The prescriptive approach from the CLU has shaped the overall approach, structure and process of community landownership, with some authors critical of the way this limits the freedom of the communities to self-organise (Wightman 2007; Brown 2008; Slee et al. 2008).

Currently over 200,000 hectares of Scotland are under community ownership (excluding community woodlands) (as shown in Figure 6.1). This is a relatively modest proportion (less than 4 per cent) of the 7.5 million hectares of rural Scotland.

COMMUNITY LANDOWNERSHIP – PANACEA OR PANDORA'S BOX FOR SUSTAINABILITY?

Community landownership is often perceived as a universal good, a logical expression of sustainable development activity which delivers widespread benefits. A review of the impacts of the Scottish Land Fund (SQW 2007) and a selection of case studies of 'social landownership' (Boyd and Reid 1999; 2000; 2001), however, highlight a range of constraints that community landowners face: economic pressures and limited income streams; limited asset bases, including a lack of affordable housing; demographic trends and continued out-migration; remoteness and inaccessibility of the purchased land; shortage of local expertise and burdens on

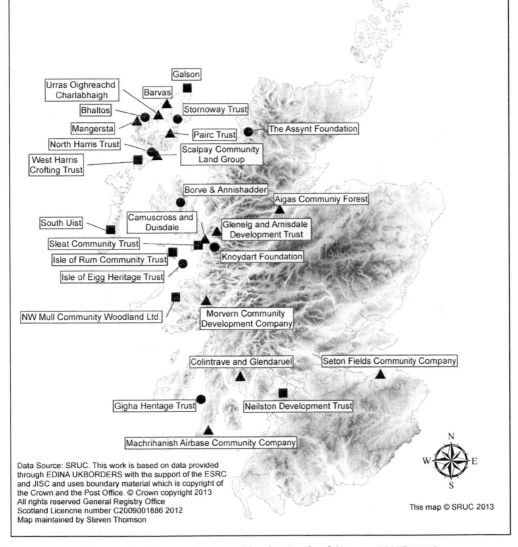

FIGURE 6.1 *Community-owned land in Scotland (Source: SRUC 2013)*

volunteers; and difficulties in achieving community cohesion. Pillai (2005) also challenges the long-term sustainability of these ventures in the light of the current economic downturn.

Despite an emphasis on economic barriers, the issue of community cohesion may also be central to the future of these initiatives: communities engaged in, or having successfully completed buyouts, are not necessarily cohesive. As studies by Brown (2008) and Rohde (2004) in Breakish and Orbost on Skye demonstrate, conflicts can erupt in conjunction with buyouts, both within communities and between communities and other stakeholders, based on conflicting values and differing definitions of what constitutes the 'community' and sustainable development. For example, in Breakish, where crofters embraced the right to buy in pursuit of a wind farm, the conflict centred around the contested identity(ies) of the community itself and whether crofters as well as non-crofting 'incomers' should have voting rights (Brown 2008). Such conflicts raise questions around the legitimacy of emerging governance structures and management committees on community land, and suggest that buyouts may also lead to the disempowerment of certain community elements compared with others.

Furthermore, the Community Right to Buy has been criticised both for failing to incorporate any clear mechanism for integrating the three pillars of sustainability and for suggesting that economic or social benefits alone are sufficient to deliver sustainable development (Pillai 2010). This raises concern over the potential environmental outcomes of community ownership given the significant scenic, natural and cultural heritage value of these areas, as revealed in the many international/national designations of the land (Warren 2009). Chenevix-Trench and Philip (2001) and Mackenzie (2006) highlight the potential for partnerships between communities and environmental organisations to deliver a more rounded approach to sustainable development and access a wider range of funding streams in a land-management context. It is here that the community as landowner becomes highly significant in re-imagining new discourses of sustainability (Mackenzie 2006, 2012). Potential for conflict between developmental and conservation-oriented objectives exists, however, both at a local scale and between community groups and their respective environmentally oriented partners (MacPhail 2002; Satsangi 2009). In particular, opposing interpretations of landscape and environment and conflicting cultural values represent key drivers of conflict (Rohde 2004). Nevertheless, the development and fostering of strategic partnerships – both formal and informal – between communities and private, public, and NGO partners, represent critical opportunities for current and future community land bodies.

Despite the challenges community landowners face, Boyd and Reid (1999; 2000; 2001), SQW (2007) and Skerratt (2011) show a range of widespread beneficial impacts accruing as a direct result of community ownership: the emergence of private enterprise; decreased out-migration because of increased security of tenure; infrastructural and rural housing projects; renewable energy schemes; enhanced community involvement; and improved land management. Furthermore, following the initial buyouts, community involvement and motivation do not appear to have been reduced, with post-buyout membership of community groups increasing in a number of cases (SQW 2007; Skerratt 2011).

A number of authors highlight the particular importance of renewable energy initiatives as a key ingredient for long-term success in community landownership (Mackenzie 2006, 2012; Slee et al. 2008; Warren 2009). The examples of Gigha and Eigg, in particular, demonstrate how community ownership can release opportunities for renewable energy, leading to financial self-sufficiency and a reversal of social and economic decline in favour of population growth and socioeconomic development (Dressler 2002; Satsangi 2009).

As well as quantifiable benefits and opportunities, community landownership also delivers a range of less tangible impacts (MacAskill 1999; Mackenzie 2004; SQW 2007; Hunter 2012). In particular, the empowerment associated with community land-ownership has been linked with increased community (and individual) confidence and belief, as well as positive community spirit and the building and release of social capital and community capacity, including leadership abilities. The experience gained by community members during and post-buyout through their direct involvement in the process appears, for example, to have a direct impact on collective community confidence and energy with regard to future buyouts (Hunter 2012). The idea that community ownership can lead to a building of capacity and associated changes in the *process* of rural development has been further developed by others (for example, Skerratt 2011).

Despite the studies discussed here, systematic assessments of the social, economic or environmental impacts of community buyouts and land reform are extremely limited (Quirk and Thake 2007; SQW 2007; Slee et al. 2008; Satsangi 2009; Skerratt 2011), with overreliance on individual case studies (Chenevix-Trench and Philip 2001; Wightman and Boyd 2001; Dressler 2002; Boyd 2003; Mackenzie et al. 2004). To address this gap, the research reported below took a 'community-based' approach to evaluation which enabled engagement with several local communities of interest, taking a strongly participatory and reflexive approach to evaluation (as suggested by Slee et al. 2008). Our case studies focused on how people experienced commu-nity ownership set within their *own* narratives of sustainability. Four key research questions emerge from the preceding review:

1. What are the individual and collective narratives of sustainable development among those directly and indirectly involved in community buyout processes and structures?
2. What are the sustainability credentials of community landownership in practice?
3. What are the key drivers of change that emerge as community ownership evolves?
4. What are the opportunities and barriers to progressing sustainability in relation to community landownership?

TABLE 6.1 Summary of key statistics for the four case studies

Case Study	Knoydart Foundation	Stòras Uibhist (South Uist)	Assynt Foundation	North Harris Trust
Year of purchase	1999	2006	2005	2003
Size (ha)	6,960	37,600	17,806	25,090
Size of community on community land	100	3,000	No resident community. Lochinver: 600.	700
No. of active crofters	1	996 crofts (850 active)	0 (some derelict crofts)	129 crofts and 22 common grazings

TABLE 6.2 Purchase costs and funding sources for the case-study buyouts

Case study	Purchase cost and funding sources
Knoydart Foundation	£750,000 [John Muir Trust (JMT), Chris Brasher Trust, Highlands and Islands Enterprise (HIE), Scottish Natural Heritage (SNH) and a neighbouring landowner]
Stòras Uibhist	£4.5 million [HIE, the Big Lottery Fund, Western Isles Council, SNH and a local appeal]
Assynt Foundation	£2.9 million [Scottish Land Fund (SLF), Tubney Charitable Trust, JMT and SNH]
North Harris Trust	£2.2 million [National Lottery's New Opportunities Fund (SLF), HIE, JMT. HIE and the SLF provided a further £250,000 to purchase the neighbouring Loch Seaforth Estate in 2006.

EXPLORING SUSTAINABILITY ON EUROPE'S WESTERN EDGE

Four examples of community landownership were selected for study within an eighteen-month time frame. These were chosen to reflect differences across key criteria: length of time since the buyout; the size of the landholding and local community; and the relative importance of crofting. All four case studies are found on the western edge of Scotland and include large areas of remote, inaccessible upland ground. Figure 6.1 shows the locations of the case studies, and Table 6.1 provides some key data about each one.

Prior to the community buyouts, the land of all four case studies was under private ownership, with the primary goals of private recreational hunting and fishing. In all four cases, purchase costs were raised from multiple sources (Table 6.2). Funding partners remain formally involved (including as board members) in a number of cases.

On all four case studies, limited companies with charitable status had been established, with membership open to local residents on the electoral roll (extending to the wider parish of Assynt in the case of the Assynt Foundation). This membership elects a board of management consisting of local residents and, in certain cases, wider stakeholders. All four limited companies have also established trading subsidiaries to engage in commercial activities. Separate renewable energy subsidiary companies have been established on Knoydart and South Uist, and a forestry subsidiary also operates on Knoydart.

Methodology: capturing different narratives of community ownership

Case-study visits were conducted between February 2010 and September 2011. Purposive sampling was employed to establish diverse interviewee groups for each case study, to allow for the capture of internal and external perspectives (Box 6.1).

Box 6.1 Interviewee types sourced in each case study (internal and external perspectives)

Chairperson and/or secretary
Development/project officer
Multiple board members (local and incomer)
Long-term and incoming residents
Local community body representatives
Neighbouring landowners
Local business owners
Crofters
Wider stakeholder representatives

Type of interviewee	Number interviewed
Staff	9 (12 per cent)
Directors	23 (30 per cent)
Subsidiary body directors	6 (8 per cent)
Wider stakeholders	22 (28 per cent)
Wider community	17 (22 per cent)

Of the total number of seventy-seven interviews, seven were carried out by phone where interviewees were not available during the visit period. Many interviewees had multiple roles: eighteen owned or co-owned their own businesses, and at least fifteen were crofters. Figure 6.2 illustrates the approach taken to capturing different narratives of how individuals experienced community ownership before, during, and after the buyout. The interview then considered the experience gained by that individual and the lessons learned. The results of this process are presented in the following sections as a series of themes, beginning with the varying interpretations of 'sustainable rural development' evident across the respondent group.

First contact with key informants
(Development managers, agency representatives
and community council representatives)

Initial exploratory visits
(Conducted on Knoydart and North Harris)

Internal and external perspectives Chairperson and/or Secretary Development/Project Officer Multiple board members (local and incomer) Long term and incoming residents Wider agency representatives Local community body representatives Neighbouring landowners Local business owners Crofters

Main site visits
(7–10 days per site – 4 sites)

Semi-structured interviews
(77 across 4 case studies)
(55 hours of recorded material)

Attending relevant events (Board meetings and conferences)

Interview themes
Drivers of buyouts,
Role and involvement of interviewee,
Perceived positive benefits and negative impacts
Constraints on sustainability;
Opportunities for the future,
Decision making, power relations and wider partnerships
Individual interpretations of sustainable rural development

**Revealing a narrative of participant experience,
experiences gained and lessons learned**

FIGURE 6.2 *Illustration of methodological sequence*

IMPACTS AND PROCESSES OF COMMUNITY LANDOWNERSHIP – NARRATIVES OF EXPERIENCE

Narratives of sustainable rural development

The majority of interviewees view sustainable rural development in relation to retention (and growth) of the local population and community capital (capacity). Three further themes were repeatedly stated to be critical: the development of participatory approaches to local governance and wider partnership working; the development of a framework to enable economic growth; and reconfiguring community–environment relationships. Table 6.3 presents these themes, which are relatively consistent across the four case studies, together with some illustrative quotes which encapsulate their meaning.

Theme 1: Community capacity (re-)building

Empowerment, capacity and the responsibilities of ownership
The community buyout process is widely seen as having created greater confidence, esteem and self-determination in case-study communities, providing a secure platform for planning the long-term future of the community. As one community worker

TABLE 6.3 Themes from an analysis of interviewee interpretations of sustainable rural development

THEMES AND SPECIFIC ELEMENTS	SELECTED QUOTES
1. COMMUNITY CAPACITY-(RE-)BUILDING (a) Long-term survival of community. (b) Retention, growth and *diversification* of local population; addressing population imbalances. (c) Developing community identity and cohesion, empowerment and capacity building; taking collective responsibility.	'That it encompasses all the elements of a viable community . . . a range of population and activities that allow it to function without excessive reliance on external inputs . . . a healthy mix of activities . . . healthy school roll, a spectrum of ages . . . a range of talent and experience' (Director). 'It's about trying to get some jobs . . . that will keep people on the island and allow them to marry and have children, which means the schools . . . and the shops stay open . . . getting to a critical mass of population, where the services will be there and people will want to live here' (Staff member). 'Nothing is sustainable unless you have people . . . you need a community where the school roll is holding its own . . . to get that you need jobs, a social environment in which people are happy to live and affordable housing. If these are right you will get children . . . the community will thrive' (Director).
2. REDEFINING PARTICIPATORY GOVERNANCE AND COLLABORATIVE WORKING (a) Participative democratic processes of governance at local community level; consensus building and leadership. (b) Collaborative development; wider partnership working. (c) Utilising local knowledge in decision making and empowering all community elements.	'It's about taking a . . . more sophisticated view of what democracy means at a community scale . . . changing the way our communities operate, share responsibilities, share problems, share opportunities. Communities . . . have a better grasp of what that [sustainability] means for their community and for their land . . . it's important for us to be judged by our family and our community, that in itself is a motivating force' (Former director). 'A sustainable community isn't one by anybody else's definition . . . they are self-defining and self-leading, that's what will make a strong community' (Director). 'It means . . . involving the whole community . . . in a meaningful way . . . that knowledge and experience base should be capitalised on . . . because that is a major resource' (Wider stakeholder).
3. BUILDING A FRAMEWORK FOR ECONOMIC GROWTH (a) Decreasing reliance on external grant funding and *running the landholding as a sustainable business.* (b) Facilitating economic growth; infrastructural development, job creation and business development. (c) Securing sustainable income streams to ensure long-term survival.	'To develop the area . . . to have a sustained income coming in to the community . . . if they don't they won't work' (Community development worker). 'To be a viable business where we get to the point where we don't have to keep going back for grants' (Community member). 'It needs to be run as a business . . . then you can deliver all of the community benefit . . . if 20 to 30 per cent of your turnover is grants . . . you will spend your time running after grants and delivering other people's expectations, if you are profitable as a business . . . you can invest in what the community wants' (Staff member).

TABLE 6.3 (continued)

THEMES AND SPECIFIC ELEMENTS	SELECTED QUOTES
4. SUSTAINABLE RESOURCE USE – ENHANCED COMMUNITY– ENVIRONMENT RELATIONSHIPS (a) Maximising the potential of assets in a sustainable way. (b) Retaining and enhancing natural assets; reworking traditional land uses; development that capitalises on natural assets; energy efficiency. (c) Increased community engagement with the land and environment; environmental land-based education/awareness	'If you take a purely . . . business-minded approach . . . you have missed out on capitalising on . . . an incredible natural environment . . . we need to make the most of the natural resources . . . not by exploiting them, but by working with them' (External stakeholder). 'It's about . . . people . . . and families living in the area . . . engaging with the land, both in terms of employment if they can . . . and using it in a way that's helping people understand and appreciate the landscape and living . . . in harmony with that' (Director).

explained, community ownership is perceived as enabling 'long-term sustainable management' and 'a permanency of focus' because 'it will always be managed for the benefit of the community'.

'I remember one lovely moment . . . when we finally got the estate and we had the wee ceilidh . . . I was getting a lift home with one of the old ladies . . . and this deer jumped out in front of the car and she put the brakes on and instead of the usual cursing . . . what [ ] who was in her seventies at that time said, big grin on her face, that's my deer now! . . . that's a huge psychological difference'. (Community worker)

This sense of empowerment post-buyout is tangible in the narratives and is a core ingredient in their reconstructions of sustainability. Some describe the 'power shift' associated with buyouts as a break away from a disempowered dependency culture towards one of empowered self-realisation and independence. This appears to engender optimism, release energy and opportunities, and stimulate interest among individuals in becoming involved with community buyouts which are often perceived as 'vehicles for change'.

Community landownership is also viewed as allowing communities fully to realise and to utilise their human potential: 'it allows access to . . . and is releasing some very creative and resourceful people'. As one staff member explains: 'It's a very mixed community here. There are the types of people who have experience and get things done and quite often – they have the right connections. I would challenge you to find a more mixed directorship in any community body.' The direct involvement of community members in all aspects of management work is building up and enhancing a local skills base in land management, business development, developing grant applications, leadership and project management. As one director explains: 'The community is developing here from being involved, their capacity is building from experience, every project they go on to they get new experiences and they pass them on to other people and it probably creates a sense of belief . . . which is a big thing.'

This capacity building is critical, with community bodies heavily dependent on volunteer support, a limited commodity in rural areas. The number of individuals available to work as directors or within working groups is often limited, with core groups commonly undertaking multiple voluntary roles in different organisations, as well as often being in full-time employment. Being involved with a community land body is viewed as a significant responsibility, requiring considerable time, as well as the capacity to absorb criticism, which often limits the numbers of people willing to stand for election to the board: 'you have to stand up and put yourself in front of eight hundred people and see how many votes you get. You have to be pretty tough and it's a big responsibility.'

As community landownership matures, continually capturing the interest of the majority (particularly younger people of working age) is therefore critical for future sustainability, with some interviewees highlighting a perceived lack of relevance of buyouts to the everyday lives of community members: 'what difference does the buyout make really, to the man on the street, it has made no difference to me'.

The initial buyout and subsequent early management are viewed as having stimulated the involvement of many of the more skilled and experienced community members. As one community director argues, a strong and continuing awareness, involvement and understanding of community governance processes are central to engendering this sense of responsibility among individuals over the long term.

Identity and cohesion

A further linked aspect of how community landownership is experienced relates to an enhanced and locally defined narrative of community development, rooted in the community itself rather than imposed from outside agencies or governments. The buyouts are viewed as 'giving a coherence to the community' and 'strengthening community identity', reuniting people in a sense of common purpose. This enhanced cohesion is most apparent in Knoydart and North Harris, with one crofter's account symbolising the importance of trust and legitimacy as key ingredients of cohesion:

> It was to do with an environmental scheme I was going into. I was trying to fence off part of my croft [and] the Commission had made a mistake. According to their record, the land wasn't in my name. The trust got the land court maps. They were completely fair to me and to the absentee tenant. That is what it was meant to be and that is why my view has completely changed of the trust.

Combined with increased confidence and capacity building, this cohesion is instrumental in facilitating wider community regeneration, encouraging investment into areas through in-migration and business start-ups. Nevertheless, demographic realities remain challenging on case-study sites with respondents from Assynt and North Harris, in particular, arguing that continued in-migration of retirees and out-migration of younger community members are indicative of a community in decline, challenging the positive factors of increased confidence and social capital outlined earlier.

Furthermore, community diversity can also fuel distinct, local 'communities of interest' and 'vocal minorities' on some community landholdings, with strongly held and sometimes opposing viewpoints potentially making community cohesion diffi-cult to achieve, at least in the short term (Box 6.2 provides an illustration). This factor is potentially exacerbated in larger, spatially diverse communities. On South Uist, for example, many of the communities now contained within the community-owned land have strong, distinct identities in their own right.

Box 6.2 'Communities of interest' and 'vocal minorities' – crofters and golf in South Uist

A golf course restoration proposal (for a course which had been in place since 1891) was approved by the then private landowner and subsequently by Stòras Uibhist in 2005; a group of seven local crofters, however, objected on the grounds that it would impinge on their grazing rights. The Scottish Land Court eventually approved the proposal on the basis that grazing continue to be allowed (a point not disputed by Stòras Uibhist). Key points include:

- Local opinion of the dispute appears mixed; local surveys show, however, that a clear majority supported the development.
- The involvement of the courts facilitated a transparent approach to conflict resolution.
- Conflicts are likely to have been associated with certain local interests having been disadvantaged from the buyout as they had 'vested interests with the previ-ous owners . . . when that's all swept away there is the perception that: (1) I am losing ground in terms of influence; and (2) these guys are . . . picking up on stuff . . . previously not bothered about' (Local resident).

Regardless of the driving forces, conflicts can be damaging for the 'public image' of community buyouts which can become 'a caricature – "the big bad landlord" – which is odd, because the "landlord" was really the community versus the "poor crofters" . . . the details get . . . lost . . . it can just become bad for the trust in terms of their public image . . . and that perception then runs through the whole thing' (Local resident).

Theme 2: Redefining participatory governance and collaborative working

Collaborative working in community landownership

Respondents from all case studies view community landownership as providing communities with a platform on which to develop new models of local participatory governance, requiring a shift from representative democracy 'to learn to work a different kind of democracy . . . by consensus if you like'. In practice, such processes are commonly associated with the idea that 'nothing will ever happen now without the community being consulted'.

As well as participatory working within buyout communities, the collaborative nature of community landownership also facilitates wider partnership working with state and non-governmental bodies, often enabling public and private investment in these areas: 'it's made the grant-giving bodies more confident about producing money . . . because they see it as a direct line into long-term community development'. Such partnerships deliver a range of tangible benefits, including infrastructure, renewable energy, housing initiatives and natural heritage. A striking example of sharing resources to achieve shared community–private objectives was evident on North Harris (see Box 6.3).

Box 6.3 Reconfiguring the role of 'community' and 'landlord' – public–private partnership on North Harris

A private businessman, who had been visiting the area for decades, was approached by the community in relation to purchasing the North Harris Estate castle and sporting rights in conjunction with the community acquiring the land and other assets. This was agreed; both facilitated the community buyout which led to the community acquiring a more cost-effective asset base. Furthermore, the businessman subsequently established a division of his international telecommunications company in an existing facility on North Harris; this now employs some twenty people.

Formal and informal partnerships are also viewed as having delivered a range of less tangible benefits, including improved working relationships. For example, a director in one case study describes how they have:

> a number of people with a track record of delivering projects, so we have credibility with a lot of these organisations [and] we can go to HIE and say 'look we have come up with a business plan against this project' and they will look upon it more favourably perhaps.

Individuals within buyout communities reveal how their involvement with partners has widened their perspectives, experience and skill sets, as well as providing them with affirmation and moral support for the work they undertake. One particularly valued outcome was the potential for sharing and accessing knowledge and information from a range of organisations, such as Highlands and Islands Enterprise, Community Energy Scotland, Scottish Natural Heritage and the John Muir Trust.

Challenges of participatory governance and partnership working
As well as delivering considerable benefits, participatory processes of governance and partnership working also pose a distinct set of challenges for community landowners.

A variety of examples of conflicting viewpoints are apparent (Boxes 6.2, 6.4, 6.5 and 6.6) highlighting differing agendas, cultural mindsets and ideologies. These differences are most pronounced in relation to development proposals and, specifically, their (apparent) potential to have an impact on the environment and/or established local interests, such as tourism or crofting. This is evident within two differing narratives which may be termed, respectively, protectionist and cultural (Box 6.4).

Box 6.4 Working 'cultural' landscape or wild land?

'Nothing annoys local people more here than to hear North-West Sutherland being described as a last great wilderness . . . it's a mismanaged landscape and it looks the way it does because of what we did to it . . . everywhere you look you see evidence of a history of people being here . . . it's not a wilderness' (Director).

'I don't see our land as wild . . . because I live in it, but also . . . I see woodlands, I see remnants of past houses from people who lived here hundreds of years ago and I see a land that has been used for thousands of years. Just because it's emptied of people doesn't mean to say it's empty . . . when you live and work somewhere you do see things differently' (Staff member).

The 'protectionist' narrative is more commonly associated with those involved in the tourism industry, environmentally oriented NGOs and public agencies, and economically inactive in-migrants in the area. 'The area is really of the highest scenic value and is protected accordingly, certain types of development . . . are simply not suitable in certain areas because they are iconic landscapes' (Wider stakeholder).

In contrast, 'cultural landscape' perspectives are more commonly associated with longer-term, locally employed community residents and those working in land-based activities. As one interviewee argues, the sense of 'belonging', which many residents feel for their area, implies a stronger motivation for community development within a longer-term perspective. Such a division of opinion, based on background and residency status, is a gross oversimplification, however, with many incoming community members heavily involved within the boards and activities of community land bodies (and supportive of a range of local developments) on all case-study estates. Nevertheless, such perspectives are revealed in relation to development proposals, such as, for example, a proposed wind farm near Suilven in Assynt (Box 6.5).

THE ROLE OF LOCAL LEADERSHIP AND COMMUNICATION IN MANAGING CONFLICT

The research also highlights the importance of 'embedded' local leadership in managing conflict and building relationships within the local community and externally. One staff member explains:

We are very lucky in having [him] as chair, because he is born and bred here, he is still a crofter and he has a position [in] the school that is respected, he speaks

---

Box 6.5 Conflicting ideologies and communication breakdowns – wind power and wild land

Following the Assynt buyout, the foundation commissioned a renewable energy feasibility study which identified a site (the Druim Suardalain ridge) within a National Scenic Area as the optimum location for two to five wind turbines. The opportunity was communicated to the community in a meeting, resulting in contentious debate and the indefinite shelving of the project. Key points include:

- Objections centred on concern among community members with tourism interests relating to landscape impacts.
- Respondents argued that long-term 'local' residents were supportive while retiring in-migrants or 'wilderness' advocates objected.
- The communication of the opportunity was agreed to have been flawed, with the presentation failing to demonstrate the 'community value' of the proposal and 'a minority of vocal objectors allowed to dominate proceedings'.
- A lack of early consultation on the proposal was viewed as having exacerbated conflict.
- Wider environmental stakeholders noted that the site was highly sensitive to development, and achieving planning permission would have been highly unlikely.
- The pursuit of such a contentious development had the potential to damage relationships with wider partners who would potentially have objected to the development.

---

the language and he knows half the population because they have been through the school. He carries an awful lot of credibility and he is anchored into the community. Sometimes you just need one figure like that – the kind of character that brings everyone together.

In-migrants can play key leadership roles in community development initiatives, however, provided there is parallel involvement of longer-term 'generational' residents. The process of securing a successful community buyout can in itself also lead to enhanced residency status, with those that had come 'through the process' becoming viewed as locally anchored leaders. The process of governance is also important, with the North Harris Trust having implemented a policy of spatial representation, with ten separate directors elected from each of the ten townships across the landholding.

A further aspect of conflict management, both within communities and between communities and wider stakeholders, relates to inclusivity and effective engagement processes. Respondents highlight how volunteer fatigue and time commitments required for major projects lead to potential misunderstandings and lack of awareness of particular decisions and actions (Box 6.6). The key lesson is the need for effective and early engagement through diverse information sources and face-to-face communication.

Box 6.6 'Goat killers' in Knoydart – the role of communication in mitigating conflict

One example of the importance of communication in minimising conflict was in Knoydart where an [initially] poorly communicated proposal by the foundation to cull the feral goat population (to lessen impacts on the vegetation) resulted in considerable debate:

> We've got wild goats . . . and we thought that locally people were aware that we were going to cull them. So I just did an update in the newsletter – that we were going to be culling x number and there was uproar. We had the press on our backs because it got mentioned [in a local newspaper] . . . it knocked out doing a cull for that year . . . We thought that we'd put information out but we obviously hadn't put enough and we hadn't discussed it enough in an open situation, because [there was] a petition . . . and we ended up with a community forum meeting. We explained what the impact was and what we were proposing and everybody said 'Oh, right, well that's OK, I wouldn't have signed it if I'd known that.' . . . We were guilty in that we hadn't got the information out and consulted . . . beforehand (Staff member).

Theme 3: Building a framework for economic development

In all case studies, economic factors are widely viewed as fundamental to the future sustainability of the communities and community land bodies. Economic aspects are more constraining in some cases than in others, however, reflecting variations in the earning potential of the estate asset base at the time of the buyout, a factor that has affected the scale and type of subsequent activities (see Table 6.4). Knoydart and Assynt were in a comparatively less secure financial position after the buyout, resulting in a greater focus on establishing stable income streams, requiring considerable investment of time and (externally sourced) finance.

In contrast, the asset base (and income capacity) on North Harris and South Uist at the point of purchase allowed (relatively) rapid development through new staff appointments and seed funding for projects. While recognising their immediate earning potential, however, some respondents also note the vulnerability of certain assets. Specific concerns relate to the long-term viability of fish farming (North Harris) which provides substantial rental income; the ongoing decline in crofting which provides rental incomes (South Uist and North Harris); the poor condition of some properties (and associated refurbishment costs) (Knoydart); and the maintenance requirements and lifespan of energy infrastructure (for example, the Knoydart HEP scheme).

The issue of financial vulnerability is a particular concern in relation to a perceived continuing reliance on public funding streams to support key posts and projects:

TABLE 6.4 Key assets in case studies at point of purchase

ASSETS	NORTH HARRIS	KNOYDART	STÒRAS UIBHIST	ASSYNT
Sporting land uses	Leased stag stalking –	Deer stalking –	Deer stalking Fishing	Deer stalking Fishing
Forestry	Limited (non-commercial)	Substantial forestry stands	–	Limited (non-commercial)
Crofts	Crofting leases	–	Crofting leases	–
Commercial property	Fish farms Hatchery	Commercial property rentals	Multiple commercial property rentals	–
Housing	–	Hostel	Grogarry Lodge	Glencanisp Lodge
Energy infrastructure and other installations	Limited – Communications Masts	Limited Hydroelectric power (HEP) scheme –	Limited – –	Limited – –

TABLE 6.5 Core activities in the four case studies

Knoydart Foundation	Stòras Uibhist	Assynt Foundation	North Harris Trust
HEP generation Ranger service Housing provision Energy efficiency improvement Sporting (deer) management	Management of croft tenancies Management/letting of Grogarry Lodge and sporting land uses (fishing /stalking) Askernish golf course development Commercial property leasing (including quarries) Major project development (6.9 MW wind energy project and harbour redevelopment at Lochboisdale)	Refurbishment of Glencanisp Lodge (sporting lodge and conference facilities) Sporting (deer/ fishing/lodge) business management	Property leasing and housing provision Renewable energy development Development of business units Management of croft tenancies

... the worrying thing about community buyouts [is that] a lot of the stuff is funded on grant aid, and the way the country is nowadays, what's going to happen when there are no grants – because you can see a lot of these things rapidly stopping quite soon without that. (Wider stakeholder)

Nevertheless, despite these challenges, interviewees repeatedly describe community landownership as a mechanism to promote local economic development set within a realisation of local assets. As one staff member pointed out, a key benefit is 'the amount of money brought in to spend ... within the community – it's hundreds of thousands which wouldn't have come in if it was a private landlord'. The core goal of such investment, as articulated by community and public body representatives, is to strengthen the financial independence of the communities in the longer term. Investment is targeted at developing the earning potential of the community asset base and increasing the potential for the emergence of new businesses, thereby allowing the communities to strengthen their own income streams, for future reinvestment in community development. Specifically within the four case studies, the focus of activity was on major infrastructural development, the development of renewable energy and affordable housing, and fostering the development of local businesses (see Table 6.5).

Major infrastructural developments
All four case studies are in relatively remote and inaccessible areas and suffer from correspondingly high building costs and long transport distances to markets. Consequently, early strategies sought to improve basic infrastructure. This included the development of a new pier and ferry terminal in Inverie (with £6 million of European, Scottish Government and Highland Council Funding), to provide an efficient, high-quality point of access to the Knoydart Peninsula. A smaller-scale (£60,000) refurbishment of a jetty has also been undertaken by the North Harris Trust. The single largest infrastructural project is on South Uist, where Stòras Uibhist is undertaking a major harbour and waterfront redevelopment initiative (Box 6.7).

Cumulatively, all these initiatives incorporate a range of measures, designed to release opportunities for socioeconomic development, increase inward private investment, and increase the viability and sustainability of the community.

Box 6.7 Releasing local economic potential – The Lochboisdale Port of Entry project

Planning approval was granted in 2011 for a major redevelopment of the Port of Lochboisdale, funded by Highlands and Islands Enterprise (£5.2 million), European Commission (£1.9 million), Western Isles Council (£625,000) and Lottery funding (£960,000). Key elements include: (a) a new 1.2-kilometre road and causeway to access over 28 hectares of land for housing and commercial developments; (b) a sailing marina; (c) a new fishing pier and dock to facilitate improvement of ferry services and docking of cruise liners. An independent assessment (Westbrook 2009) predicted nineteen full-time equivalent (FTE) jobs in construction, with a further ninety-one FTE jobs from the development of businesses and support in the ten years following construction, giving an additional annual household income locally of £2.07 million.

Green energy as an economic lifeline
The development of renewable energy initiatives is seen as one of the most important opportunities for securing future income streams. Interviewees perceive renewable energy within a 'win-win' narrative, recognising its contribution to climate-change mitigation, while also funding local community regeneration. Indeed, potential income from renewable energy was an important element of the financial justification for a community buyout in all case studies. With the exception of Knoydart (Box 6.8), however, no existing renewable energy schemes were included with the community buyout.

Box 6.8 Powering an isolated community – The Knoydart Hydro Scheme

At the time of the Knoydart buyout, an existing HEP scheme (the main local electricity source) was in severe disrepair. A £500,000 refurbishment programme began in 2001 (funded by HIE, European Commission, and the foundation). The scheme employs two part-time staff and provides power to over seventy homes and businesses, with income being reinvested in further improvements (£40,000 invested in 2009–10) or used to pay off loans used to fund the initial refurbishment. As electricity demand is increasing (by 30 per cent in 2009–10), the scheme represents a source of future income.

Initiatives vary in scale, with Stòras Uibhist currently engaged in developing the largest community-owned renewable energy installation in Scotland, having shelved an earlier 'business renewables' initiative because of changes in government incentives. The projected income from the Lochcarnon initiative (Box 6.9) is envisioned as the economic engine of Stòras Uibhist's plans for local development. On Assynt and North Harris, early proposals to install three to five wind turbines failed, because of local and wider stakeholder objections in Assynt (Box 6.5), and excessive turbulence on the Harris site. On North Harris, smaller-scale initiatives have been pursued to establish further income streams (see Box 6.10). As well as renewable energy developments, various examples of energy-saving and low-carbon initiatives are in place, including an energy-efficiency position in Knoydart and the insulation of buildings on North Harris.

Box 6.9 Scotland's largest wholly community-owned renewable energy initiative

Stòras Uibhist secured funding (£2.4 million from the European Commission, £1 million from Social Investment Scotland, £8 million in loans from the Co-operative Bank) and planning permission to install three 2.3 MW wind turbines on Lochdar Hill Common Grazings at Lochcarnon. Projected electricity generation equates to an annual net income to Stòras Uibhist of £1.5 million, with two-thirds of this used for loan repayments in the first ten to twelve years (after which the total £1.5 million will be available). Income will be used to support community development and new local businesses.

Box 6.10 A mixed bag of renewable energy on North Harris

The North Harris Trust had secured planning permission (2010) and grid connection (2011) for a 150 kW hydroelectricity scheme. Projected development costs are £638,000, with an income-generation capacity of £100,000 per year. Loan funding was provisionally secured in 2011; an ongoing British government review of the feed-in tariff (FIT), however, resulted in funding being put on hold for many HEP schemes (including Bunavoneader and two other micro-hydro schemes on North Harris). A 10 kW wind turbine has also been developed at the Recycling Centre which powers the site and generates annual income of £5,000 to £6,000. Two 5 kW turbines have also been commissioned for the local youth club, and planning permission has been granted for three 5 kW turbines to reduce school fuel costs and power a future community building and offices.

There has therefore been limited success thus far in relation to renewable energy initiatives, with the Lochcarnon initiative indicative of the major potential in terms of income. In contrast, proposals have failed or been delayed in three of the four case studies, all of which have involved considerable time and cost input in developing proposals, and failed or stalled for different reasons: changes to government incentive schemes; turbulence; and community and wider stakeholder resistance relating to perceived landscape and/or environmental impacts.

Affordable housing

A lack of affordable housing is a critical issue in all case studies and commonly seen as a key driver of out-migration and community decline. Different approaches have been taken to tackle this issue across the case studies: (a) refurbishing existing properties (Knoydart and North Harris); (b) developing new properties, usually in partnership with other organisations (Knoydart and North Harris); (c) selling housing plots to stimulate private self-builds (Knoydart, North Harris, Stòras Uibhist); (d) developing housing opportunities in conjunction with other new developments (North Harris, Stòras Uibhist – see Box 6.6); (e) developing housing policies (North Harris and Knoydart) and acquiring rural housing body status (Knoydart); and (f) exploring options for reletting vacant crofts (Stòras Uibhist).

On both Knoydart (three properties with further projects ongoing) and North Harris (two properties) a number of properties have been refurbished and made available for rent. New properties have also been developed in both these cases, with the largest affordable housing development (with eight completed homes and three serviced plots) completed as a partnership initiative on North Harris in 2011 (see Box 6.11 and Figure 6.3). New builds have also been completed in Knoydart, including three A-frame buildings and three new family homes through the Scottish Government's Rural Homes for Rent scheme. This has been the only successful application to the scheme from a community landowner.

Box 6.11 Working together for affordable housing on North Harris

Through consultations and discussions with local crofters, the North Harris Trust identified an opportunity for an affordable housing initiative in Bunavoneader. Land was leased from Ardhasaig Common Grazings and a partnership established between Hebridean Housing Partnership (HHP), Tighean Innse Gall (TIG), North Harris Trust and Ardhasaig Grazings. Construction was completed in 2011, with eight affordable homes established to let and three fully serviced plots for purchase.

Selling off plots for private housing development has also occurred, with plots for sale on North Harris and Knoydart. On Knoydart, site sales have occurred in conjunction with the development of a shared equity scheme, with the foundation retaining a 20 to 25 per cent share of any sold property (land and/or buildings). This system makes property more affordable, as well as ensuring that a proportion of all sold assets remains within the community, allowing the foundation control over their assets and ensuring sold properties remain in permanent occupancy. Two affordable apartments have also been developed in conjunction with the development of new offices for the North Harris Trust.

Fostering local business development and job creation

Despite some success stories, the development of new private businesses is limited across the case studies, given the short time since the buyouts and, perhaps more critically, a lack of existing infrastructure to facilitate the emergence of local businesses,

FIGURE 6.3 *Affordable homes on North Harris*

including a lack of pre-developed business units and supporting services, such as high-speed broadband.

In response, targeted development of facilities to encourage business 'start-ups' has either occurred or is at some stage of development in all case studies. On North Harris, a proposal by the trust to develop zero-carbon business units has reached the design phase (see Box 6.12) while, on Knoydart, the foundation directly assisted a local community member in establishing a business through developing facilities (see Box 6.13); a similar initiative is evident in Assynt. Progress is also being made on Knoydart and North Harris in relation to improving the speed and consistency of local broadband services.

Box 6.12 Providing a space for innovation and business start-ups on North Harris

In 2010, the North Harris Trust commissioned a feasibility study exploring the potential for development of zero-carbon business units. Three possible sites were identified and initial designs were developed which are adaptable to different business types, built from stone and timber and powered by renewable energy. Full design funding has now been secured, and planning and construction funding is being sought. The trust plans to advertise the units, using their zero-carbon credentials and the high quality of life, to attract three new businesses to the area.

Box 6.13 Releasing home-grown entrepreneurism and retaining the community in Knoydart

One community member explains how the Knoydart Foundation played a role in her decision to return to the community and establish a business. While studying pottery at college, options were explored for establishing a local pottery business and café. The foundation renovated a property and, once the community member had completed her training, she established a successful business with another community member. The initiative provided career opportunities for those involved, which led to the retention of younger community members, while increasing the value of the community asset base.

Some respondents argue that out-migration has led to the loss of many of the more entrepreneurial community members resulting in an uncertain dependency on in-migration for new business development. A minority of respondents argue that this is reflected in the priorities of their respective community land bodies, through a perceived weak focus on business development and economic activity in favour of community- and environmentally oriented projects.

Nevertheless, existing and emerging businesses are benefiting directly and indirectly from the activities of the four community land bodies. In particular, local construction and machinery/haulage companies recognise that they derive income from involvement in infrastructural and construction projects stimulated and/or led by community land bodies. Existing businesses, including quarries, fish farms and craft-based businesses, also often represent an important source of income to community land bodies through rental of land or buildings. On Knoydart and North Harris, business owners also note how community ownership has raised the profile of their area, with associated benefits for those involved in tourism.

Projects developed by community land bodies also provide an alternative avenue of work for self-employed professionals in areas where opportunities are often limited. This 'spin-off' activity also has impacts for the long-term unemployed, with one council-developed training and employment scheme on Harris benefiting directly from involvement in North Harris Trust activities through the emergence of a diverse set of projects for them to work on.

Community land bodies have also employed people directly in a range of skilled posts (Table 6.6), and developed businesses and income streams in their own right, usually through a trading subsidiary. Businesses and initiatives developed within, or directly linked with, community land bodies vary, with sporting businesses in three case studies, renewable energy initiatives present or emerging in three, and accommodation provision in all four (Table 6.7).

Theme 4: Reconfiguring community–natural resource relationships

In all four case studies, community landownership is perceived as having resulted in a reconfiguring of resource use and land management away from a previously

TABLE 6.6 Numbers directly employed and type of posts within community land bodies

	Assynt Foundation	North Harris Trust	Knoydart Foundation	Stòras Uibhist
Development manager/chief executive	1	1	1	3
Development officer	1			1
Renewable energy and energy-efficiency projects		1	1	
Land management		1		
Rangering		1	1	
Community engagement		1		
Administration/office management		2	1	2
Supporting local development officer post			1	
Crofting officer				1
TOTAL POSTS	2	7	5	7

TABLE 6.7 Businesses and initiatives generating employment developed within, or linked to, community land bodies

Case study	Main businesses and income streams managed, developed or under development by community land bodies	No. of posts
Assynt	Commercial deer stalking and fishing	1
	Glencanisp Lodge (accommodation and events facility)	2
North Harris Trust	Community Recycling Centre	1
	Micro-renewables (potential post, under development)	1
Knoydart Foundation	Commercial deer stalking and fishing	2
	Knoydart Foundation Hostel	1
	Hydro power provisioning (Knoydart Renewables)	2
	Knoydart Forest Trust (linked community-based initiative)	6
Stòras Uibhist	Grogarry Lodge	3
	Fishing and sporting	2
	Askernish golf course (directly linked)	1
	Lochcarnon renewable energy	1
	Port of Entry (project management and future port management)	2
Total		25

sporting estate-centred, passive, preservationist approach, towards more proactive, community-centred approaches. Community landownership is often perceived as a mechanism for reconnecting rural communities with the land and the environment, shown through a range of specific measures relating to land management, environmental interpretation, renewable energy development and energy efficiency.

(Re-)Working the land

On all case studies, crofting is generally viewed as being in decline (owing to changes in agricultural support and out-migration) and dependent on crofters having other sources of income and/or amalgamating multiple crofts to increase efficiency. Where crofting is more prevalent (North Harris and South Uist), measures had been taken to administer this land use more efficiently and to support crofters (for example, through advising on grant applications). Stòras Uibhist had also improved drainage systems, consulted on measures to create opportunities for new entrants, and carried out coastal protection work (with Scottish Natural Heritage) to stabilise dune systems and protect coastal crofting areas.

Crofting is largely absent from Assynt and Knoydart, despite aspirational objectives to re-establish crofts, with some respondents highly critical of this lack of prioritisation of the re-establishment of this land use. In contrast, both these case studies have more prevalent forestry- and woodland-expansion objectives, with a major forest-restructuring programme being undertaken in Knoydart by the Knoydart Forest Trust, with the aim of creating a functioning forest ecosystem.

Deer stalking is commonly viewed as having been symbolic of the exclusivity of private ownership, and the emphasis with respect to deer management has changed in all four case studies. Following the buyouts, deer culling, deer population counts, vegetation monitoring, and deer management planning had been implemented (or were in the process of being developed) in line with conservation objectives and designated site requirements. For economic reasons, however, deer stalking and other sporting activities (such as fishing) remain a source of income in all cases except North Harris. This has resulted in the general aim of maintaining deer densities which facilitate both commercial deer stalking and habitat conservation.

This is challenging in practice and, in certain case studies, environmental stakeholders have concerns relating to the lack of emphasis immediately after the buyout on increasing deer culling and more timely development of agreed deer management plans, despite management agreements and funding stipulating such measures within a certain time frame. Particularly in one case study, certain environmental stakeholders express similar concerns in relation to a clash of perspectives on wider environmental issues:

> . . . it became quite clear that we could not at that time stay as part of the board of directors of an organisation like this . . . there was a conflict of interests really and things had become difficult . . . there was a view that we were a sort of form of opposition if you like.

Nevertheless, it is widely agreed that perceptions of sporting management have evolved, now being viewed on community landholdings primarily as a source of income and a component of a wider land-management strategy. As one external stakeholder explains:

> if you are not involved in it, you see stalking as a rich upper-class sport. I see stalking as a business that allows us to bring in income that allows us to implement our deer management and vegetation management policies.

This is reflected further in initiatives being developed on Knoydart, Assynt and South Uist to market venison locally, to create further income and integrate local venison into the community.

One of the more striking examples of changing relationships between local communities and natural resources is the development of a community stalking group on North Harris (see Box 6.14) and the development of low-cost measures to allow locals to fish lochs and rivers in Harris and South Uist. This has led to the decline of poaching and changing local perceptions of stalking and fishing activities.

Box 6.14 Community-based deer management on North Harris

On North Harris, as part of the buyout agreement, the fishing rights and rights to stalk thirty stags were purchased separately, together with the castle, by a private individual; the remaining stalking rights were obtained by the community, with a community stalking club established to manage the remaining deer. This was facilitated initially by the (private) gamekeepers taking out club members, with members subsequently stalking unsupervised. The club is widely supported, though one respondent remains sceptical of a community-based approach to deer management 'How can fifty people manage deer . . . you have got fifty people going out all wanting a good hind . . . that can't be good for your herd.'

Key aspects of this initiative include:

- The club has twenty-six members and is open for membership (for a £5 annual fee and a payment of £10 per hind) to people from North Harris and certain outlying areas.
- Members must have deer stalking and firearms certificates, insurance and a game licence.
- The group collectively carries out deer-population counts, agrees cull targets, and undertakes and records culling.
- The approach negates the need to employ a deer manager and functions as a barrier to illegal deer stalking, limiting the potential for future conflicts.
- The club represents a tangible example of how community landownership can deliver benefits for individuals, an important factor for maintaining long-term interest in the Trust.

Interpreting and communicating the environment
As well as direct land-management measures, community land bodies also engage in interpretation and communication of the environment. This is particularly apparent on North Harris and Knoydart because of the employment of rangers who engage in a range of activities (see Table 6.8). Rangers are viewed as playing a key role in providing local people (including children) and visitors with opportunities to experience local environments and engendering a sense of responsibility for the land.

Enhancing environmental awareness and opportunities for experiencing wildlife and high-quality landscapes are also linked with potential ecotourism opportunities, such as wildlife-watching holidays. The North Harris Trust has developed an eagle-watching hide to encourage such opportunities, as well as managing a substantial upland path-restoration initiative, while the Knoydart Foundation offers opportunities

TABLE 6.8 Activities carried out by rangers on Knoydart and North Harris

Education and interpretation	• Environmental education and activities programme for local schools • Development of interpretive materials • Leading guided walks, wildlife watching and photography expeditions • Co-ordinating Duke of Edinburgh/JMT award expeditions/activities
Conservation and habitats	• Clearance of non-native species • Wildlife monitoring and development of wildlife watching areas
Other activities	• Organising conservation volunteer groups • Team-building activities • Forestry activities; path maintenance; fencing; and deer counts • Campsite development and maintenance

for 'photo-stalking' of deer. As many interviewees argue, the local economy depends strongly on visitors and this, in turn, depends on the maintenance of a high-quality natural environment which necessitates that the local community takes a sustainable approach. This perspective positions the environment as an integral part of the asset base, lessening potential tensions between community and environmental interests.

These views further indicate the move away from passive–preservationist approaches, towards proactive, sustainable land management which secures the sustainability of the environment and, in so doing, secures the future of the community.

DISCUSSION

Moving towards a sustainable future

These findings, set within the context of the Scottish Highlands and Islands, reveal community landownership functioning as a powerful catalyst and positive agent for reconstructing rural development set within locally prescribed narratives of sustainability. A collective post-ownership 'identity', which maximises the potential for managing change to best advantage, is emerging: both the process and the outcomes combine within a reinvigoration of community cohesion, pride and ambition. Such processes were not without dispute and conflict, however, both within and across different communities of interest. While, traditionally, conflict may be seen as problematic, this forms a necessary phase of sustainability discourse and reinvigoration. Conflict management and effective leadership therefore represent critical aspects of future long-term progress towards sustainability.

Indeed, these findings reinforce much of the wider literature and case studies with respect to the social and economic impacts of community ownership (for example, MacAskill 1999; Boyd and Reid 1999, 2000, 2001; Mackenzie 2004; SQW 2007). The discussion below, however, focuses on the more interpersonal aspects, as the case studies provide a wealth of insight that adds significantly to existing work.

Specifically, the analysis shows a collective increase in confidence and ambition among local communities and wider stakeholders uniting around a locally conceived and constructed vision of sustainability that prioritises investment in infrastructural

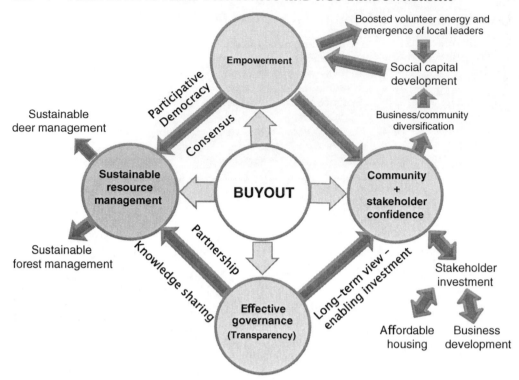

FIGURE 6.4 *The 'sustainability cascade' effect as conceptualised on the basis of case-study analysis*

and housing initiatives, supporting local entrepreneurial activity and reconstructing community–environment relationships. Set within a rapid learning environment, there has been a concomitant increase in skills and experience which has further empowered and energised communities, leading to the emergence of local leaders. The establishment and wider legitimacy of new local governance structures has further enhanced community capacity and secured opportunities for partnership working – facilitating knowledge and resource sharing which, when combined with participatory approaches, appears to be leading to more sustainable resource-management outcomes. Collectively, a cascade effect becomes apparent with respect to rural development processes, which is conceptualised in Figure 6.4.

These process-related impacts echo those recorded by Skerratt (2011) who concluded that the building of both material (for example, investment in assets) and social benefits (for example, increased confidence), which is occurring on community-owned land, combine to enhance community resilience (the capacity of a community to adapt to change – see Magis 2010).

Investing in social capital

Social capital is a core ingredient in Themes 1 (community capacity rebuilding) and 2 (participatory governance and collaborative working) and Boxes 6.3, 6.11 and 6.14.

Lee et al. (2005) and Ray (1998) argue that social capital is constructed and enhanced through processes of 'political struggle' and strongly linked with the development of a single and unified sense of identity in a rural area. Set within each of the case studies, the community ownership process has involved significant struggles and conflict management, as befits any new form of governance. These experiences and processes have resulted directly in a strengthening of community identity and increased community resilience. As well as the development of localised (place-specific) social capital, wider partnerships also represent a critical aspect for success, with a range of associated outcomes, including social learning and resource sharing (Theme 2, Boxes 6.3 and 6.11). Relationship building between homogeneous groups, however, can also be counterproductive for wider community cohesion (Nelson et al. 2003; Sanginga et al. 2007). This may at least partly explain the conflicts evident in the case studies, particularly those between geographically disparate or sectoral groupings, such as on South Uist (further reported on by Bryan 2011) and Assynt. Shucksmith (2000), for example, describes how formal crofting networks are often male dominated and exclude non-crofters, potentially resulting in social exclusion. Thus, social capital is not a universal good; latent social capital within different groups may result in conflicting viewpoints and entrenched positions during periods of change, such as during and following a buyout, or in response to a development proposal. The vocalisation of certain community elements and resulting negative perceptions of 'conflict' may therefore be a relatively natural (and even necessary) outcome of consensus-based participatory democratic processes.

Managing conflict

The occurrence of conflict in these cases (for example, Boxes 6.2 and 6.5), and as reported elsewhere (for example, Rohde 2004; Brown 2008; Bryan 2011), can have a negative impact on community energy and capacity, and necessitates careful consideration of the role, structure, and composition of governing bodies within community land contexts. Conflict and its management, however, represent core and necessary components of community development activity. These findings suggest that processes of geographic representation (which were firmly in place on North Harris), wherein directors are elected on a parish basis, offer potential for managing conflict across large rural areas.

Conflicting views between community bodies and wider stakeholders (for example, Box 6.5; Macphail 2002) may, in certain cases, relate to a tension between, on the one hand, a desire to promote local participation and capacity building and, on the other, a reliance on a set of managerial systems (such as auditing or management-plan templates) to ensure institutions are accountable (Mackinnon 2002). Such tensions suggest a need for the development of adaptive, place-specific approaches and/or carefully formulated concordats prior to the establishment of partnerships.

These findings also demonstrate the critical importance of communication and inclusive consultation as a cornerstone of conflict management. Pre-emptive communication and direct involvement (for example, through working groups) in processes and initiatives are also fundamental to ensuring long-term majority community involvement.

The perceived impacts, both negative and positive, of in-migrants or 'white settlers' on rural demographics and cohesion have been observed in these case studies and more widely (for example, Jedrej and Nuttall 1996). As Burnett (1998) argues, however, despite development rhetoric presenting the 'local' as a known concept, the category is not established in concrete terms and, in reality, Scottish rural development generally embraces the incomer as a source of capital. This view is supported here within the concept of 'local embeddedness' – being more related to direct involvement in the buyout and being known locally – rather than having a generational link to the locality. Such embedded leaders are critical in developing consensus-based approaches both locally and with respect to developing bridging capital and accessing powerful networks.

Economic challenges and the importance of the community asset base

As Slee et al. (2008) and Thake (2006) note, compared with many specialist and/or long-standing (private) landowners, many community land bodies are undercapitalised and (at least initially) may lack financial and management expertise. The differing economic positions relate primarily to the income-generating capacity of the respective asset bases at, or immediately following, the point of purchase. This factor has been critical in terms of communities being able to acquire funding, projects and staff. The time required to develop sustainable income streams, particularly on asset-poor and access-limited sites, should not be underestimated. Furthermore, the asset bases of future aspirational buyouts should be very carefully assessed in terms of earning potential. As apparent from Figure 6.5, which illustrates the links between the challenges faced by community landowners, a limited asset base not only limits economic viability but also directly affects the potential for community growth and the development of social capital and cohesion.

Renewable energy represents a considerable opportunity to develop income in certain cases; barriers exist, however, including a reliance on costly expertise and (changeable) government incentive schemes, and potential resistance to proposals because of perceived landscape impacts. As a result (see Theme 3), the economic development potential of renewable energy identified in pre-buyout feasibility studies is not always achievable. Emergent 'eco-economic' activities (such as ecotourism, wood fuel and venison), which merge place and new production–consumption supply chains, may also offer potential to capitalise on limited but high-quality resources and emerging technologies and market places (Marsden 2011).

As a result, in locations where assets are limited, a longer-term commitment of state support may often be required, to ensure that communities have the capacity to develop their own income streams and become financially self-reliant over the longer term. Critically, community landownership is a relatively new phenomenon in land-management and community-development terms; any accurate judgement of economic sustainability must therefore be made over the longer term. Furthermore,

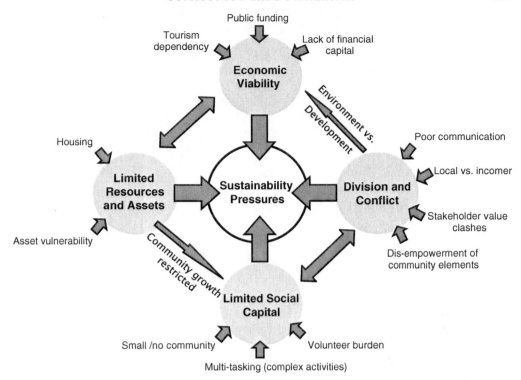

FIGURE 6.5 *A summary of the 'sustainability pressures' faced by communities on case-study estates*

it should be noted that community buyouts are not the only beneficiaries of state support, with the top fifty recipients of agricultural subsidy in Scotland alone (mainly large private landowners) receiving a total of £168 million in the 2000–9 period (an average of £3.3 million per landowner) (Scottish Government statistics quoted in Wightman 2011).

In all the case studies, tourism represented a significant income stream, recognising the value of environmental assets, but reworked into a local discourse of sustainability. Significantly, under changed governance structures, there is increased interest in conservation designations in order to help secure new funding avenues for rural development, with the North Harris community, for example, proposing that their area be designated as a national park. Such changes are profound, given the historical traditions of resisting imposed designations (Scott and Shannon 2007).

Reconstructing sustainability

These findings concur with the view that land-reform processes lack clarity on sustainable development outcomes (Pillai 2010) but provide an important setting wherein new narratives of sustainability can emerge. These are generally community centred and place-specific and not always agreed by all stakeholders, with variable perceptions of acceptable trade-offs between sustainability strands (see Box 6.5).

Local development objectives may, therefore, drive future conflict between different interests on community land holdings, based on the various state, organisational and community-based interpretations of sustainability. A counterpoint, however, as explored by Evans and Birchenough (2001), is that community management of natural resources will, by default, be more sustainable, because of communities having higher levels of motivation, knowledge and experience relating to resources that are inextricably bound up with their livelihoods. Furthermore, partnerships, which are often necessary to acquire funding and support, offer the potential to widen sustainability perspectives (Chenevix-Trench and Philip 2001). These findings reflect those of Mackenzie (2006, 2012) and Satsangi (2009) who argue that community ownership (of North Harris and Gigha respectively) has led to a local reworking of concepts of nature and sustainability, away from the idea of a 'preserved wilderness' towards one of the 'working wild' where active engagement with, and sustainable consumption of, the environment play key roles in local development. Developing and applying coherent sustainability frameworks which are sufficiently flexible to facilitate endogenous development, while satisfying state and wider protocols and objectives, therefore represents a key future challenge for community landowners (and consideration can be given to the applicability of the sustainability framework presented in Chapter 8 in this context). Critically, such processes require an adaptive approach which ensures continued capacity building, continued community involvement and builds momentum for the future.

References

Boyd, G. (1999). 'To restore the land to the people and the people to the land'. In: Boyd, G. and Reid, D. (eds), *Social landownership*, Volume One: *Eight Case Studies from the Highlands and Islands of Scotland.* Community Learning Scotland, Inverness. 80 pp.

Boyd, G. (2003). '"What Evidence' Said the One-eyed Public Policy Maker?" A Brief Review of Recent Community Land Ownership Trends in Scotland'. <http://www.caledonia.org.uk/land/evidence.htm> (last accessed 1 June 2009).

Boyd, G. and Reid, D. (eds) (1999). *Social Landownership*, Volume One: *Social landownership – Eight case studies from the Not for Profit Landowners Project in the Highlands and Islands of Scotland.* Caledonia Centre for Social Development. <http://www.caledonia.org.uk/social-land/nfp.htm> (last accessed 21 September 2012).

Boyd, G. and Reid, D. (eds) (2000). *Social Landownership*, Volume Two: *Social landownership – Eight case studies from the Not for Profit Landowners Project in the Highlands and Islands of Scotland.* Caledonia Centre for Social Development. <http://www.caledonia.org.uk/social-land/nfp.htm> (last accessed 21 September 2012).

Boyd, G. and Reid, D. (eds) (2001). *Social Landownership*, Volume Three: *Social landownership – Eight case studies from the Not for Profit Landowners Project in the Highlands and Islands of Scotland.* Caledonia Centre for Social Development. <http://www.caledonia.org.uk/social-land/nfp.htm> (last accessed 21 September 2012).

Brennan, M. (2001). 'Melness Crofters Estate, Sutherland "A new crofting community landowner"'. In: Boyd, G. and Reid, D. (eds), *Social Landownership*, Volume Three: *Social landownership – Eight case studies from the Not for Profit Landowners Project in the Highlands and Islands of Scotland.* Caledonia Centre for Social Development. <http://www.caledonia.org.uk/socialland/nfp.htm> (last accessed 21 September 2012).

Brown, A. P. (2008). 'Crofter forestry, land reform and the ideology of community'. *Social and Legal Studies* 17 (3), pp. 333–49.

Bryan, A. (2011). 'Stòras Uibhist 2006–2010 Independent Review, Final Report, April 2011'. Commissioned report for Highlands and Islands Enterprise and Stòras Uibhist.

Bryden, J. and Geisler, C. (2007). 'Community-based land reform: Lessons from Scotland'. *Land Use Policy* 24, pp. 24–34.

Burnett, K. A. (1998). 'Local heroics: reflecting on incomers and local rural development discourses in Scotland'. *Sociologia Ruralis* 38 (2), pp. 204–24.

Chenevix-Trench, H. and Philip, L. (2001). 'Community and conservation land ownership in highland Scotland: A common focus in a changing context'. *Scottish Geographical Journal* 117 (2), pp. 139–56.

Dressler, C. (2002). 'Taking charge on Eigg – The benefits of community ownership'. *ECOS* 23 (1), pp. 11–17.

Evans, S. M and Birchenough, A. C. (2001). 'Community-based management of the environment: lessons from the past and options for the future'. *Aquatic Conservation* 11, pp. 137–47.

Hunter, J. (2012). *From the low tide of the sea to the highest mountain tops: Community ownership of land in the Highlands and Islands of Scotland.* Islands Book Trust, Lochs. 204 pp.

Jedrej, M. C. and Nuttall, M. (1996). *White settlers: the impact of rural repopulation on Scotland.* Harwood, Luxembourg. 200 pp.

Lee, J., Arnason, A., Nightingale, A. and Shucksmith, M. (2005). 'Networking: social capital and identities in European rural development'. *Sociologia Ruralis* 45 (4), pp. 269–83.

MacAskill, J. (1999). *We have won the land.* Acair, Stornoway. 224 pp.

Mackenzie, A. F. D. (2004). 'Re-imagining the land, North Sutherland, Scotland'. *Journal of Rural Studies* 20 (3), pp. 273–87.

Mackenzie, A. F. D. (2006). '"S Leinn Fhein am Fearann" (The land is ours): re-claiming land, re-creating community, North Harris, Outer Hebrides, Scotland'. *Environment and Planning D* 24, pp. 577–98.

Mackenzie, A. F. D. (2012). *Places of Possibility: Property, Nature and Community Land Ownership.* Wiley-Blackwell, London. 270 pp.

Mackenzie, A. F. D., MacAskill, J., Munro, G. and Seki, E. (2004). 'Contesting land, creating community, in the Highlands and Islands, Scotland'. *Scottish Geographical Journal* 120 (3), pp. 159–80.

MacKinnon, D. (2002). 'Rural Governance and local involvement: assessing state–community relations in the Scottish Highlands'. *Journal of Rural Studies* 18 (3), pp. 307–24.

MacPhail, I. (2002). 'Relating to land: the Assynt Crofters Trust'. *ECOS* 23 (1), pp. 26–35.

Magis, K. (2010). 'Community Resilience: An Indicator of Social Sustainability'. *Society & Natural Resources* 23 (5), pp. 401–16.

Marsden, T. (2011). 'Mobilities, vulnerabilities and sustainabilities: exploring pathways from denial to sustainable rural development'. *Sociologia Ruralis* 49 (2), pp. 113–31.

Nelson, B. J., Kaboolian, L. and Carver, K. A. (2003). *The Concord Handbook: How to Build Social Capital Across Communities.* The Concord project, Los Angeles.

Pillai, A. (2005). 'Community land ownership in Scotland: progress towards sustainable development of rural communities?' In: Thompson, D. B. A., Price, M. F. and Galbraith, C. A. (eds), *Mountains of Northern Europe: Conservation, Management, People and Nature.* TSO Scotland, Edinburgh, pp. 295–8.

Pillai, A. (2010). 'Sustainable Rural Communities? A legal perspective on the community right to buy'. *Land Use Policy* 27, pp. 898–905.

Quirk, B. and Thake, S. (2007). 'Making assets work'. The Quirk review of community management and ownership of public assets. HMSO, London.

Ray, C. (1998). 'Culture, intellectual property and territorial rural development'. *Sociologia Ruralis* 38, pp. 3–19.

Rohde, R. (2004). 'Ideology, Bureaucracy and Aesthetics: Landscape Change and Land Reform in Northwest Scotland'. *Environmental Values* 13, pp. 199–221.

Sanginga, P. C., Kamugisha, R. N. and Martin, A. M. (2007). 'The dynamics of social capital and conflict management in multiple resource regimes: a case study of the Southwestern Highlands of Uganda'. *Ecology and Society* 12 (1), pp. 6.

Satsangi, M. (2009). 'Community Land Ownership, Housing and Sustainable Rural Communities'. *Planning Practice and Research* 24 (2), pp. 251–62.

Scott, A. J. and Shannon, P. (2007). 'Planning for rural development in Scotland: A new role for local landscape designations?' *Planning Theory and Practice* 8 (4), pp. 509–28.

Shucksmith, M. (2000). 'Endogenous development, social capital and social inclusion: Perspectives from LEADER in the UK'. *Sociologia Ruralis* 40, pp. 208–18.

Skerratt, S. (2011). 'Community landownership and community resilience'. Rural Policy Centre Research Report, Scottish Agricultural College, Edinburgh. <http://www.sac.ac.uk/mainrep/pdfs/commlandownerfulllowres.pdf> (last accessed 8 January 2013).

Slee, B., Blackstock, K., Brown, K., Dilley, R., Cook, P., Grieve, J. and Moxey, A. (2008). *Monitoring and Evaluating the Effects of Land Reform on Rural Scotland – a scoping study and impact assessment*. Scottish Government Social Research, Edinburgh. 210 pp.

SQW (2005). 'Evaluation of the Community Land Unit'. Final report to Highland and Islands Enterprise, Inverness.

SQW (2007). 'The Scottish Land Fund Evaluation'. Final Report to the Big Lottery Fund. <http://www.biglotteryfund.org.uk/er_eval_slf_final_report.pdf> (last accessed 8 January 2013).

Thake, S. (2006). *Community Assets: the Benefits and Costs of Community Management and Ownership*. Department of Communities and Local Government, HMSO.

Warren, C. (2009). *Managing Scotland's Environment*. Second Edition. Edinburgh University Press, Edinburgh. 432 pp.

Wightman, A. (1997). *Who owns Scotland?* Canongate, Edinburgh, 237 pp.

Wightman, A. (2007). 'Land Reform (Scotland) Act 2003. Part 2, The community right to buy. A Two Year Review'. Caledonian briefing No. 6, Caledonian Centre for Social Development.

Wightman, A. (2011). *The Poor Had No Lawyers: Who owns Scotland and how they got it*. Reprint Edition. Birlinn, Edinburgh. 320 pp.

Wightman A. and Boyd, G. (2001). 'Not-for-Profit Landowning Organisations in the Highlands and Islands of Scotland: Organisational Profiles and Sector Review'. Report to HIE and SNH.

CHAPTER SEVEN

Buying nature: a review of environmental NGO landownership

Robert Mc Morran and Jayne Glass

INTRODUCTION

Since the 1980s, there has been a reinvigorated movement by environmental non-governmental organisations (NGOs) to purchase land in the United Kingdom, including estates in the uplands of Scotland. This has occurred in direct response to a perceived failure of government conservation policies, including the nature conservation designation system, and has taken place in the context of growth in public environmental awareness and in the membership of environmental NGOs (Aitken 1997; Wightman 2000; Chenevix-Trench and Phillip 2001). Direct purchase of land has often occurred in response to perceived threats from inappropriate development in order to secure permanent protection of the natural heritage and/or landscape features and move from advocacy to practical demonstration of conservation management practices.

As a group, charitable conservation organisations have significant landholdings in Scotland. The National Trust for Scotland (NTS) is the third largest NGO landowner (with 78,000 hectares); the Royal Society for the Protection of Birds (RSPB) owns almost 50,000 hectares; and the John Muir Trust (JMT) 24,461 hectares (Warren 2009; JMT 2012; NTS 2012a). As well as facilitating the protection and enhancement of landscapes, habitats and wildlife, ownership of land provides these organisations with opportunities to explore innovative approaches to land use and management, and to use their land as demonstration sites to influence wider land-management practices (Johnston 2000; Chenevix-Trench 2004).

Globally, environmental NGOs have been subject to criticisms regarding: the lack of formal definition of their roles (Austin and Eder 2007); a lack of accountability to the people they claim to represent (Simmons 1998; Lehman 2007); a lack of awareness and understanding of local knowledge in the places they work (Levine 2002; Sælemyr 2004; Gordon 2006; Johnston and Soulsby 2006); and, when designing and implementing projects, the use of large amounts of their own resources in ways that overlook local existing capacities and responsibilities (Collier 1996). This chapter provides a critique of environmental NGO landownership, based primarily on a review of academic and wider literature, both from Scotland and further afield. Particular consideration is given to examples of how these organisations manage their larger landholdings in Scotland's uplands, drawing on a range of secondary data sources.

ENVIRONMENTAL NGOs – THE WIDER CONTEXT

NGOs play key intermediary roles between governments and the private sector, and have become increasingly important in facilitating the transfer of technology, promoting public awareness, and negotiating on international environmental issues (Simmons 1998; Moon and Park 2004). Globally, the involvement of NGOs in land management is often expected to promote a 'bottom-up' approach and self-reliance (Ito et al. 2005), with NGOs regularly viewed as institutional representatives of civil society (Levine 2002). In this regard, international donor agencies have given renewed prominence to the role of NGOs (Mohan 2002), commonly viewed as 'vehicles of social change' addressing challenges resulting from the inability of governments to address environmental problems (Price 1994, 42). Since the 1970s, the number of NGOs operating at regional or national scales has increased; most are united under the (international) rubric of sustainable development (Price 1994).

Direct land purchase by environmental NGOs for the purposes of conservation, as opposed to taking an advisory or partnership role, is not a new phenomenon (Pierce 1996; Fairfax et al. 2005). Indeed, the full or partial acquisition of land has been described as the 'predominant focus of terrestrial conservation strategies' (Davies et al. 2010, 29). For example, the Dutch NGO, Natuurmonumenten, has bought numerous nature reserves in the Netherlands; in 2012, it had 732,000 members (Knegtering et al. 2002; Natuurmonumenten 2012). Funding from membership fees and public appeals are of particular importance for funding land purchases (and NGOs generally) in developed countries, with international donors of greater importance in developing countries.

Equally notable is a growing preference in a range of mainly industrialised countries to employ conservation easements (or covenants) to develop voluntary but legally binding agreements between landholders and environmental NGOs where the covenant remains on the title of the property even if it is sold (see Rissman et al. 2007; Green Balance 2008; Morris 2008). Voluntary certification is also growing in importance, with the European Pan Parks initiative (supported by World Wildlife Fund and the Dutch travel company Molecaten), for example, utilising a wilderness certification scheme to raise awareness of European wilderness areas and promote sustainable and responsible tourism practices through collaboration between government bodies, NGOs and tourism operators (Pan Parks 2011). Partnership-based approaches, which may or may not involve land purchase depending on the specific local context, are also becoming more prevalent. Rewilding Europe, for example, aims to establish ten high-quality wildlife and wilderness areas by 2020 through large-scale ecological restoration, with initiatives developed through a collaboration between funding partners, local conservation partners and rewilding partners (for example, expert advisers) (Rewilding Europe 2012).

The rise of conservation NGO landownership in Scotland

In Scotland, several environmental NGOs have purchased land for conservation purposes (see Box 7.1). These landholdings, many purchased in recent decades, include numerous upland sites of high conservation and recreational value (Aitken 1997; Boyd and Reid 1998; Chenevix-Trench and Philip 2001; Croft 2004). These organisations have a variety of management objectives for the sites they manage, linked to their organisational aims and objectives (Maxwell and Birnie 2005). Site management is often heavily supported by public funding (Maxwell and Birnie 2005), with wider organisational development also supported financially and otherwise by large memberships (Chenevix-Trench 2004), of whom very few – if any – are resident on the land in question (Arnott 1997) or even in Scotland (for example, most RSPB members).

The growth of estate ownership in Scotland by environmental NGOs has been attributed to a range of factors, including: safeguarding outstanding scenery and landscapes for current and future generations; protecting areas for amenity use (access); protecting wildlife and habitats; perceived threats from inappropriate development and the failure of government environmental and access policy, and protecting and restoring wild land (Wightman 1996; Boyd and Reid 1998, 2000; Maxwell and Birnie 2005; Warren 2009). Conservation groups have been described as supplementing environmental legislation through the ownership of land (Chenevix-Trench and Philip 2001), and it has been suggested that 'the great advantage of conservation ownership is that it can provide continuity and the organisation involved may feel better able than a sporting landowner to look outside the crude subsidy system in deciding how to develop the land' (Cramb 1996, 73). In line with wider sustainability agendas, the objectives of environmental NGOs have become increasingly integrated, recognising the importance of wider society and local communities when managing land for conservation (Box 7.1), though it has been questioned whether such policies are put into practice widely (Keating and Stevenson 2006).

Box 7.1 Environmental NGOs owning significant areas of land in upland Scotland (Sources: Featherstone 2004; Ashmole and Ashmole 2009; Warren 2009; Wightman 2010; SWT 2011; JMT 2012; NTS 2012a; WTS 2012)

The **National Trust for Scotland** (NTS), established in 1931, owns and manages 78,000 hectares of countryside properties, including sixteen islands. The trust has four core purposes: conservation, access, education and enjoyment. The trust's larger and remoter rural properties are also managed according to a Wild Land Policy, based on maintaining the wildness of the landscape to ensure continuity of high-quality recreational experiences and scenic landscapes.

The **Royal Society for the Protection of Birds (RSPB)** (established in 1889) manages 66,562 hectares across Scotland, spread over more than seventy-six reserves (49,948 hectares are owned outright, 7,541 hectares are leased and 5,486 hectares are managed via management agreements). The organisation has the core aim of conserving and enhancing wild bird populations and the environments on which they depend to enrich the lives of people and ensure the long-term maintenance of the ecosystems upon which human life depends.

The **John Muir Trust** (JMT), established in 1983, is Britain's 'leading wild land charity' and owns and manages 24,461 hectares of the wildest parts of Scotland. The trust's vision is that: wild land is enhanced; wild land is protected; people engage with wild places; and communities thrive alongside wild land. As well as managing land through direct acquisition, the trust works in partnership with other landowners to achieve its objectives, including four community land trusts and one private landowner. Crofting also occurs on some of the trust's properties.

The **Scottish Wildlife Trust** (SWT) acquired its first property in 1966 and now owns and manages 120 nature reserves across Scotland totalling 20,800 hectares. Many sites are urban or peri-urban, given the focus of the trust on inspiring and engaging people in wildlife and conservation. There are also sites in the Highlands and Islands, including on Eigg and Handa and in Assynt (Ben Mor Coigach). The core aim of the trust is to establish a network of resilient ecosystems supporting expanding communities of native species across large areas of Scotland's land, water and seas.

Woodland Trust Scotland acquired its first property in 1984 and now owns 8,750 hectares of woodland across Scotland on eighty sites from Stranraer in the south to Sutherland in the far north. The trust aims to work with others to plant more native trees; to protect native woods and trees and their wildlife for the future; and inspire everyone to enjoy and value woods and trees.

Trees for Life (TFL) was established in 1989 with the core aim of restoring a wild forest in the heart of the Scottish Highlands, for its own sake, as a home for wildlife, and to fulfil the ecological functions necessary for the well-being of the land itself. TFL has a three-way strategy of encouraging natural regeneration of native woodlands, planting native trees and removing non-native tree species, and works in partnership with organisations such as Forestry Commission Scotland to achieve its aims.

The **Borders Forest Trust** was formed in 1996 with the aim of restoring native woodlands and other natural habitats in southern Scotland. The trust works in partnership with local people, communities and organisations to connect people with woodlands in an effort to re-establish a woodland culture in the south of Scotland.

The purchase of land by these organisations is not a recent trend; the NTS acquired Glencoe over several stages between 1935 and 1937 and now owns 5,680 hectares of the glen (NTS 2005). These purchases took place as a result of a growing sense of public grievance and resentment following the nineteenth-century Trespass Act and Game Laws which restricted public access on estates in Britain (Boyd and Reid 1998). The increased tension between hillwalkers/climbers and sporting estates led to a nationwide appeal in the 1930s which culminated in the purchase of the estate, with significant support from the Scottish Mountaineering Club and Percy Unna,

TABLE 7.1 Notable events in the history and growth of voluntary conservation organisations in the UK in the twentieth century (Sources: Wightman 1996; Adams 2003; Boyd and Reid 1998; Wightman 2000; Warren 2009)

1930s	A limited number of voluntary conservation organisations existed: Council for the Preservation of Rural England (CPRE) and equivalents in Scotland and Wales, The Royal Society of Wildlife Trusts, National Trust, National Trust for Scotland and the RSPB.
1940s–50s	Sector began to grow with the establishment of county naturalists trusts.
1958	Council for Nature created to lobby on behalf of the conservation organisations.
1960–65	Number of wildlife trust members grew from 3,000 to 21,000. Scottish Wildlife Trust formed in 1964.
1961	World Wildlife Fund (WWF) established.
1971	Friends of the Earth (FoE) established in 1969, becoming an international network in 1971, with national FoE organisations established in England in 1971 and in Scotland in 1978.
1977	Greenpeace UK established.
1970s–1980s	Membership and income of all organisations grew (e.g. RSPB had over 10,000 members in 1960 and exceeded 500,000 in 1980; membership of the National Trust, NTS and Ramblers' Association doubled over the 1980s). JMT established in 1983, Trees for Life established in 1989.
	A number of large, voluntary conservation organisations acquire property important for wildlife and scenic landscapes (RSPB, JMT, SWT).
1988	Wildlife trusts across the UK had an annual income of £6–7 million and held and managed over 1,700 nature reserves, covering 50,000 hectares.
1980–95	Total area of land owned by NGO sector in Scotland rose by 146 per cent to reach 133,500 hectares. Scottish Countryside LINK (now Scottish Environment LINK) established (1980).
1997	Membership of RSPB reaches one million, with a total income of over £32 million.
2000 onwards	JMT membership passes 10,000 (2009). Wildlife trusts across the United Kingdom have 800,000 members and manage over 2,000 reserves. Total land under conservation NGO ownership surpasses 185,000 hectares. Net income of RSPB in 2011 is £89.3 million. In 2013, Scottish Environment LINK has thirty-three organisations as members.

then president of the NTS (Boyd and Reid 1998). Unna's preferences for future management of the land became known as the 'Unna Principles' which emphasised the maintenance of the 'primitive' values of the land under the ownership of the trust, and have since been enshrined in the Wild Land Policy of the NTS (NTS 2002).

Following World War II, the membership of voluntary conservation organisations was quite limited because of low morale and weak leadership (Adams 2003). A steady growth in support and public concern for the countryside (Wightman 2000), however, as well as increased public willingness to pay for the protection of species and habitats (Croft 2004), culminated in a marked increase in the membership of these organisations towards the end of the twentieth century (see Table 7.1). This gave them considerable bargaining power and increasing influence in the political arena (Arnott 1997; Chenevix-Trench and Philip 2001; Adams 2003).

Scottish land under NGO ownership includes islands, iconic and remote mountain landscapes, heavily designated nature reserves, and some of the largest and most valuable areas of semi-natural habitats (for example, Caledonian pinewoods) in Britain. The NTS, for example, owns 128 properties and has been the third largest landowner in Scotland since the purchase of the 29,380 hectare Mar Lodge Estate in 1995 (Huband 2004). The estate lies in the heart of the Cairngorms National Park and its management adopts an integrated approach, combining traditional sporting activities, large-scale native woodland restoration, access provision, and landscape enhancement. The RSPB has also become a major landowner, owning and managing 66,562 hectares across Scotland in 2007, a 31 per cent increase since 2000 (Warren 2009). The JMT owns eight properties, including several iconic peaks such as Ben Nevis (bought in 2000 for £450,000), Schiehallion (1999) and Quinag (2005). The Scottish Wildlife Trust (SWT) and the Woodland Trust Scotland (WTS) have also purchased land. Table 7.2 details some key upland estates owned by these organisations, including information about location and management practices.

ENVIRONMENTAL NGO LANDOWNERS – PART OF A SUSTAINABLE FUTURE FOR SCOTLAND?

The sustainability policy of the Scottish Government suggests that voluntary organisations have an important part to play in the delivery of 'a more sustainable Scotland' (Scottish Executive 2005, 90). In the context of estate ownership and management, conservation organisations therefore have a considerable opportunity to demonstrate good practice in order to increase understanding of the value and benefits of a well-stewarded natural environment. A number of key themes can be identified in relation to the impacts and benefits of environmental NGO landownership from a sustainability perspective. These include: (a) protection and restoration of native species and semi-natural habitats, including data collection and expansion of the information base relating to species and habitats; (b) the protection and restoration of a range of iconic landscapes; (c) local socioeconomic impacts, including increased local spend and wider involvement of people with the natural environment; (d) partnerships and collaborative working, including the provision of advice and collaborative landscape-scale conservation initiatives and increased political advocacy on environmental issues. This section explores these themes in further detail.

Species and habitat conservation

A key focus of the management of most large rural properties owned by NGOs is the protection and/or reintroduction of native species and large-scale restoration of habitats and ecosystems. An emphasis on native woodland restoration is particularly evident, with the RSPB at Abernethy, for example, attempting a large-scale restoration of native Caledonian pinewood through deer culling to facilitate natural woodland regeneration (Beaumont et al. 2005). This reflects the approach taken by Scottish Natural Heritage at Creag Meagaidh (Putman et al. 2008), and reductions in deer population are increasingly being adopted as a mechanism to allow native woodland regeneration on other sites (see Box 7.2).

TABLE 7.2 Some key estates owned by NGOs in upland Scotland (Sources: Wightman 1996; Adams 2003; Boyd and Reid 1998; Wightman 2000; Warren 2009)

NGO	Estate	Size (ha)	Date of acquisition	Management practices and other comments
NATIONAL TRUST FOR SCOTLAND	Mar Lodge, Aberdeen-shire	29,340	1995	Sport shooting; long-term ecological restoration; deer management; access management and restoration; tourism; holiday accommodation.
	Glencoe, Lochaber	5,800	1935	Conservation; sustainable tourism practice and demonstration; visitor interpretation and education; deer management.
	St Kilda	854	1957	Conservation; enjoyment and education and managed as a model of integrated conservation management. Designated as a National Nature Reserve and World Heritage Site.
ROYAL SOCIETY FOR THE PROTECTION OF BIRDS	Abernethy Estate	12,406	1988	Forest regeneration; tourism (visitor centre at Loch Garten); habitat enhancement; deer management; volunteer projects; education.
JOHN MUIR TRUST	Skye estates: Torrin, Sconser, Strathaird	12,044	1991–97	Regeneration of grazings; monitoring; conifer restructuring; access management and path restoration; crofting; management of cultural heritage; volunteer groups.
	Ben Nevis	1,761	2000	Preserving the mountain's wild character; education of visitors and maintaining access while minimising erosion and negative impacts.
	East Schiehallion	871	1999	Path restoration; visitor management and education; habitat management and restoration.
SCOTTISH WILDLIFE TRUST	Ben Mor Coigach	6,198	1998	Species and habitats conservation; forest management; access management; crofting; livestock grazing.
WOODLAND TRUST SCOTLAND	Glen Finglas	4,085	1996	Woodland regeneration and planting; conifer restructuring; education; access management and trail provision.
TREES FOR LIFE	Dundreggan Estate	4,046	2008	Caledonian forest restoration; ecological monitoring and research; volunteer conservation projects and group management; education.
BORDERS FOREST TRUST	Carrifran	650	2000	Native woodland restoration; tree planting; education and research; access and recreation. Initial purchase part funded by six hundred individual founders, each providing a minimum of £250.

> **Box 7.2** Re-establishing a wild heart of the Highlands –
> Dundreggan Estate, Trees for Life[1]
>
> The long-term ambition of Trees for Life (TFL) is the restoration of some 1,600 square kilometres of Caledonian pinewood, re-establishing a 'wild heart' of the Highlands in the form of a fully functioning forest ecosystem, including (reintroduced) keystone predators such as lynx and wolf. The strategy of TFL has included the purchase of the 4,000 hectare Dundreggan Estate in Glen Moriston in 2008, as well as wider partnership working with organisations such as Forestry Commission Scotland, to achieve their vision. The purchase of Dundreggan has facilitated partnership working through a liaison group with neighbouring landowners and the local community.
>
> The long-term vision for 2058 is to see Dundreggan restored to a wild landscape of diverse natural forest cover, with the return of species such as red squirrel, capercaillie, golden eagle, European beaver and wild boar. Managing deer to reduce grazing pressure is a management priority, and steps are being taken to renaturalise conifer plantations, restore mires, and ensure that any new planting does not take place on deep peat. A small part of the estate is adjacent to the River Moriston, and this is being managed to create a large, riparian aspen zone, in a joint project with Forestry Commission Scotland. Trees for Life generates some income from venison sales from culling activities on the estate, and hosts visitors on volunteer conservation holidays in estate accommodation.

The conservation of peatlands is also continuing on a number of sites, including the RSPB's 5,305 hectare Forsinard reserve (see Box 7.3) in the Flow Country (Sutherland) and Plantlife's 1,237 hectare Munsary reserve in Caithness.

> **Box 7.3** Peatland restoration on Forsinard reserve (Source: RSPB 2010a)
>
> On Forsinard reserve (established in 1995), the RSPB, in partnership with Scottish Natural Heritage and Forestry Commission Scotland, have engaged in blanket bog restoration at a landscape scale, funded by the European LIFE Programme. Drains have been blocked across 15,600 hectares of blanket bog, using over thirteen thousand dams, and planted trees have been removed from 2,200 hectares of former blanket bog. Rewetting of drained areas is facilitating the recovery of bog mosses and other vegetation, raising water levels and allowing peat to grow. Collectively, these efforts have also attracted wading birds, such as golden plover and greenshank. A research centre has been established on the site, and is engaged in long-term monitoring and is becoming (through wider collaborations) a major centre for research on peatland ecology, hydrology and carbon. An interpretation centre has also been established on the reserve, attracting some four thousand visitors a year who contribute some £190,000 to the local economy.

Conservation of wetland and grassland habitats, through the maintenance of traditional grazing regimes, is also practised on some sites, including the RSPB's Insh Marshes reserve in the Cairngorms. As well as habitat conservation, NGOs, including the SWT and RSPB, are involved in species conservation and/or reintroduction programmes, often in collaboration with government agencies, including initiatives relating to red squirrel, capercaillie, golden eagle, and European beaver (Scottish Squirrel Group 2004; Whitfield et al. 2008; FCS 2008; SWT 2010). Landscape-scale habitat restoration and species-based approaches are not without their controversies, with deer population reductions, in particular, sometimes resulting in conflicts with surrounding landowners, such as on the Mar Lodge estate (Windmill et al. 2011). In addition, representatives of the private landowning sector have described species reintroductions and 'rewilding' as unsustainable and unsuitable in cultural and in ecological terms in a Scottish context (for example, Cooke 2012).

Landscape protection and enhancement

As well as habitat- and species-based approaches, an emphasis on the protection and enhancement of landscapes is apparent among NGO landowners. In addition to the maintenance of cultural landscapes, such as through the maintenance of grazing regimes, there is a clear focus on wild land. In particular, on wild land, both the NTS and JMT have specific policies which are centred on limiting the degree of human influence on remote and scenic landscapes (Box 7.4).

Box 7.4 Selected definitions of 'wild land'

Scottish Natural Heritage – Wildness in Scotland's countryside (SNH 2002): The term 'wild land' is best reserved for those limited core areas of mountain and moorland and remote coast, which mostly lie beyond contemporary human artefacts such as roads or other development.
National Trust for Scotland – Wild land policy (NTS 2002): Wild land in Scotland is relatively remote and inaccessible, not noticeably affected by contemporary human activity, and offers high-quality opportunities to escape from the pressures of everyday living and to find physical and spiritual refreshment.
John Muir Trust – Wild land policy (JMT 2010): Uninhabited land containing minimal evidence of human activity.

The JMT has an explicit focus on the protection and enhancement of wild land and this reflects a growing interest nationally in the concept: the authorities responsible for both of Scotland's national parks have mapped wildness across their respective park areas in recent years (Carver et al. 2012), and Scottish Natural Heritage followed a similar approach to map wildness across Scotland in 2012 (SNH 2012a). As well as minimising developments regarded as intrusive, such as renewable energy infrastructure and vehicular hill tracks, the protection and enhancement of wild land has also incorporated measures such as fence removal, the 'feathering' of linear conifer plantation edges, and the restoration of eroded footpaths (Mc Morran et al. 2006). Taking a broader perspective (to include ecological land-management issues) the JMT has developed a set of management standards for wild land (Box 7.5).

Box 7.5 A set of Wild Land Management Standards

In an effort to encourage land managers across Scotland to apply an ecosystem approach and enhance the wildness of the land they manage, and to apply a consistent approach across their own sites, the John Muir Trust launched a website in 2012 that presents a set of management standards for wild land (www.wildlandmanagement.org.uk). The site includes a management template and a Wild Land Management Standards handbook, and details twenty-eight management standards falling into six categories: management planning; soil, carbon, water; biodiversity and woodland; deer and livestock; facilities and heritage; communities, visitors and awareness.

Socioeconomic benefits and impacts

The ownership and management of land by NGOs delivers a range of social and economic benefits, including the development of environmental interpretation, opportunities for recreational experiences in urban/peri-urban and wilder sites, increased local visitor numbers and local spend, local employment, and the involvement of people in land management and conservation through volunteering opportunities. The development of interpretive facilities by NGOs is significant, with sites such as the NTS Glencoe Visitor Centre and the RSPB's Loch Garten Osprey Centre attracting some 120,000 to 150,000 and thirty thousand to forty thousand annual visitors respectively (Bryden et al. 2010; SNH 2012b). A number of sites also incorporate wildlife-watching infrastructure, such as the capercaillie-watching hides at Abernethy, enhancing opportunities for wildlife experiences and photography.

As well as providing opportunities for 'wild' experiences through limiting large-scale developments and enhancing habitats, NGOs also maintain and improve access, particularly by improving and/or constructing footpath networks. The JMT, for example, raised over £800,000 to fund major realignment and restoration work on the path to the summit of Schiehallion (on a popular mountain estate in Perthshire owned by the trust) between 1999 and 2003 (Glass 2010). The NTS, which maintains eighty-two high-level routes on seven mountain properties, established the Mountain Heritage Programme in 2003, resulting in £1.9 million being spent on upland footpath repairs between 2003 and 2009 (SNH 2009). Their more recent 'Mountains for People' programme will run for four years and cost £1.25 million (Jones 2011). Improved access and wider awareness of key sites influence visitor numbers, with annual visitors to the NTS's Mar Lodge estate (which has a 210 kilometre footpath network) and the RSPB's Abernethy Forest reserve increasing in recent years to over a hundred thousand and seventy thousand respectively (Taylor 2007; NTS 2012b). In many cases, some visitors act as part-time staff, with volunteer workers often assisting with land-management activities on NTS, RSPB, SWT and JMT sites.

Direct spending by NGOs on site management and related activities, together with direct employment, increased visitor numbers and associated local spend, can all have positive impacts on local economies. Direct spend by the RSPB across five of their key Scottish reserves (excluding staff costs) totalled over £450,000 in 2002 (Shiel et al. 2002). Estimated direct spend over five years at Abernethy alone was £645,000 (Shiel et al. 2002), with eleven people employed directly in site management and in the visitor centre, and a further estimated seventy-six jobs sustained locally through the associated increase in visitor numbers (Beaumont et al. 2005). A more recent study concluded that RSPB reserves across the whole of Britain attracted £66 million to local economies in 2009 and supported 1,872 full-time equivalent jobs (over 50 per cent of which were linked with tourism-related activities), and that the local employment associated with reserves had increased by 87 per cent since 2002 (Molloy et al. 2011).

More generally, scenery and wildlife both represent important tourist attractions in a Scottish context (VisitScotland 2008), with nature-based tourism currently worth some £1.4 billion annually to the Scottish economy, supporting 39,000 full-time jobs (Bryden et al. 2010). Environmental NGOs therefore have an important role to play in maintaining, as well as enhancing, this resource. Furthermore, Mc Morran et al. (2006) concluded that the management of land for wildness values represents a competitive alternative to traditional land uses; Taylor (2007) goes further, suggesting that local economic benefits of 'rewilding' projects (not all of which are on NGO-owned sites), including the number of full-time jobs sustained, are up to five times greater than those of traditional land uses.

While estates purchased and managed primarily for the conservation or improvement of habitats and/or wild land can bring economic and employment benefits to local economies, however, conservation organisations in Scotland do not generally acquire and manage these estates with the primary aim of producing such benefits, preferring to measure their success largely in terms of their own conservation objectives. Consequently – and not simply within the context of estate ownership – there have been accounts of hostility towards nature conservationists who have 'descended on local communities [. . .] and told them what they had to do with their land' (Holdgate 2003: 57). This has led to certain organisations becoming unpopular among the communities who live and work on the land that these organisations own. Additionally, conservation organisations have been criticised for conceptualising the countryside as a giant nature reserve or wilderness where wildlife protection is deemed more important than traditional land uses and the quality of rural livelihoods (Mitchell 1999). Certain landowning NGOs have even been described as 'absentee landlords', with all the pejorative connotations of a term that is more typically associated with privately owned estates (Huband 2004). This is caused by a widespread perception that management decisions are often 'remote' and 'top down' in nature, made in the headquarters of the respective organisation, and giving insufficient weight to local concerns. Such negative feelings can, in some cases, result in the alienation of local communities (Maxwell and Birnie 2005) and create conflict: on the JMT-owned

Sandwood Estate, for example, crofters at one time refused to serve on a committee with non-crofters (Chenevix-Trench and Philip 2001). These negative impressions and conflicts flow from perceptions that conservation organisations are detached from (and insensitive to) the local community and local knowledge. In many respects, the long-term success of nature conservation initiatives (and sustainable management more generally) depends on partnership working and building consensus at both local and other scales, as discussed below.

Collaborative working

Participatory and collaborative approaches represent a cornerstone of many emergent conservation initiatives in Scotland. Landscape-scale habitat and ecosystem restoration initiatives, such as the Futurescapes initiatives led by the RSPB, the Living Landscapes projects led by the SWT, and the Great Trossachs Forest initiative, involve a range of stakeholders, including all of the major landowning NGOs, public bodies, community landowners and private landholdings (Hughes and Brooks 2009; RSPB 2010b; Scottish Forest Alliance 2012). These initiatives aim to restore landscapes and habitats through re-establishing ecological functions and connectivity at regional scales, in conjunction with nature-based socioeconomic development. As well as collaborating with other NGOs and public agencies, some NGOs also collaborate directly with community groups; for example, JMT has been active as a partner on a number of community buyouts (see Chapter 6), provides advice to private estates on conservation management, and has involved local communities in the development of estate-management plans. Such collaborative approaches offer considerable scope for sharing knowledge, resources and responsibility. Whether such initiatives can both generate income and provide a basis for local economic development, however, thus leading to longer-term sustainability, remains an open question.

DISCUSSION – NGO LANDOWNERSHIP IN TWENTY-FIRST-CENTURY SCOTLAND

Conservation organisations have been described as having a 'unique opportunity to demonstrate to the people of Scotland and beyond that sound conservation management can live alongside, and indeed help to promote, economic development for the wider benefit of remote areas' (Croft 2004, 81). The notion of moving from advocacy to practical demonstration ('doing what governments cannot or will not do') is common among conservation NGOs that own land. There is, therefore, an opportunity to realise that estate ownership also brings the possibility for innovation which is arguably where NGOs may have more potential to add value than other types of owner by drawing on skills, expertise, funding, and research to which other owners may not have access. Furthermore, there is considerable potential for landowning NGOs to act as facilitators, supporting and enabling communities to achieve sustainable development goals, recognising the potential for conservation to be a means to an end, rather than an end in itself. In many respects, this strengthening of the integration of conservation goals with local community development goals represents one of the greatest challenges for landowning NGOs in Scotland.

Questions also remain with regard to how far into the future conservation organisations are looking, as well as the extent to which ownership by environmental NGOs can have a long-term, financially independent future. It is a challenge for NGO-owned estates to be financially self-sufficient, just as it is for privately owned estates. In the mid 2000s, for example, running Mar Lodge Estate cost the NTS over £600,000 annually (Warren 2009). Whether NGOs should continue to buy land remains an open question; it is arguable whether they should instead focus their efforts on exerting more influence on other landowners to encourage conservation-focused practices through strategic partnerships, the development of demonstration sites, and the provision of advice. For example, staff from the JMT work in partnership with Corrour Estate in Lochaber, a privately owned estate to which they provide conservation management consultancy. Potential also exists for NGOs to function as 'first-aid organisations' – as opposed to long-term landowners – that move their resources around for maximum impact. In this manner, they could 'ask themselves not how much more land they can buy, but how soon they can get rid of it', passing estates on to communities in the long run (Wightman 2000, 183). Ultimately, there remains scope to assess the extent to which the ownership of estates by NGOs delivers sustainability goals in upland areas, in particular asking whether charitable ownership models which focus primarily on environmental and conservation management satisfy the public interest. For now, though, conservation ownership is well established as a significant and influential part of the picture of landownership and management in upland Scotland.

NOTE

1. This case study illustration is drawn from the work presented in Chapter 8, which piloted a sustainability tool on some estates owned by environmental NGOs.

REFERENCES

Adams, W. M. (1996). *Future Nature: a vision for conservation*. Second Edition. Earthscan, London. 294 pp.

Aitken, R. (1997). 'A vision for Scotland's finest landscapes'. In: *Protecting Scotland's Finest Landscapes: Time for action on National Parks*. Conference proceedings, 17 September 1997. Scottish Wildlife and Countryside Link, Perth, pp. 9–14.

Arnott, J. (1997). 'The protection of the land'. In: Magnusson, M. and White, G. (eds), *The Nature of Scotland: Landscape, Wildlife and People*. Canongate, Edinburgh, pp. 211–25.

Ashmole, P. and Ashmole, M. (2009). *The Carrifran Wildwood Story: Ecological Restoration from the Grass Roots*. Borders Forest Trust, Jedburgh. 224 pp.

Austin, R. L. and Eder, J. F. (2007). 'Environmentalism, Development, and Participation on Palawan Island, Philippines'. *Society and Natural Resources* 20 (4), pp. 363–71.

Beaumont, D. J., Amphlett, A. and Housden, S. D. (2005). 'Abernethy Forest RSPB Nature Reserve: managing for birds, biodiversity and people'. In: Thompson, D. B. A., Price, M. F. and Galbraith, C. A. (eds), *Mountains of Northern Europe: Conservation, Management, People and Nature*. TSO Scotland, Edinburgh, pp. 239–50.

Bond, I., Davis, C., Nott, K. and Stuart-Hill, G. (2006). 'Community based natural resource management manual'. Wildlife Management Series, WWF for a Living Planet. <http://awsassets.panda.org/downloads/cbnrm_manual.pdf> (last accessed 8 January 2013).

Boyd, G. and Reid, D. (eds) (1998). *Social Land Ownership*, Volume One, *Eight more case studies from the Highlands and Islands of Scotland*. The Not-for-Profit Landowners' Group, Inverness.

Boyd, G. and Reid, D. (eds) (2000). *Social Land Ownership*, Volume Two, *Eight more case studies from the Highlands and Islands of Scotland*. Community Learning Scotland, Inverness.

Bryden, D. M., Westbrook, S. R., Burns, B., Taylor, W. A., and Anderson, S. (2010). *Assessing the economic impacts of nature based tourism in Scotland*. Scottish Natural Heritage Commissioned Report No. 398.

Carver, S., Comber, A., Mc Morran, R. and Nutter, S. (2012). 'A GIS Model for mapping spatial patterns and distribution of wild land in Scotland'. *Landscape and Urban Planning* 104 (3–4), pp. 395–409.

Chenevix-Trench, H. and Philip, L. J. (2001). 'Community and Conservation Land Ownership in Highland Scotland: A Common Focus in a Changing Context'. *Scottish Geographical Journal* 117 (2), pp. 139–56.

Chenevix-Trench, H. (2004). 'Conservation land ownership in Scotland: time to pause for thought?' *ECOS* 25 (2), pp. 37–43.

Collier, C. (1996). 'NGOs, the Poor and Local Government'. *Development in Practice* 6 (3), pp. 244–9.

Cooke, R. (2012). 'Chris Packham lynx proposition well wide of the mark'. *Scope: The Newsletter of the Association of Deer Management Groups* 3 (25), p. 10.

Cramb, A. (1996). *Who owns Scotland now? The use and abuse of private land*. Mainstream, Edinburgh. 206 pp.

Croft, T. A. (2004). 'Conservation Charity Land Ownership in Scotland'. *Scottish Geographical Journal* 120, pp. 71–82.

Davies, Z. G., Kareiva, P. and Armsworth, P. R. (2010). 'Temporal patterns in the size of conservation land transactions'. *Conservation Letters* 3, pp. 29–37.

Fairfax, S. K., Gwin, L., King, M. A., Raymond, L. and Watt, L. A. (2005). *Buying Nature: the limits of land acquisition as a conservation strategy, 1780–2004*. The MIT Press, Cambridge, MA. 360 pp.

Featherstone, A. W. (2004). 'Rewilding in the north-central Highlands: an update'. *ECOS* 25 (3/4), pp. 4–10.

Forestry Commission Scotland (2008). 'Action for Capercaillie'. <http://www.forestry.gov.uk/pdf/fcs-action-capercaillie.pdf/$file/fcs-action-capercaillie.pdf> (last accessed 8 January 2013).

Glass, J. H. (2010). 'Access to the hills: five years on'. *Munro Society Journal* 2, pp. 67–72.

Gordon, J. E. (2006). 'The role of science in NGO mediated conservation: insights from a biodiversity hotspot in Mexico'. *Environmental Science and Policy* 9, pp. 547–54.

Green Balance (2008). 'The Potential of Conservation Covenants'. A report by Green Balance to the National Trust. August 2008.

Holdgate, M. (2003). 'The human stake in nature'. *ECOS* 24 (1), pp. 57–62.

Huband, S. (2004). 'Social responsibility and the National Trust for Scotland; time to trust the Trust'. *ECOS* 25 (2), pp. 31–6.

Hughes, J. and Brooks, S. (2009). 'Living landscapes: towards ecosystem-based conservation in Scotland'. Scottish Wildlife Trust, Edinburgh.

Ito, K., Oura, Y., Takeya, H., Hattori, S., Kitagawa, K., Paudel, D. and Paudel, G. (2005). 'The influence of NGO involvement on local people's perception of forest management: a case study of community forestry in Nepal'. *Journal of Forest Research* 10, pp. 453–63.

John Muir Trust (2010). 'Wild Land Policy'. John Muir Trust, Pitlochry.

John Muir Trust (2012). 'John Muir Trust Properties'. <http://www.jmt.org/properties.asp> (last accessed 15 December 2012).

Johnston, E. and Soulsby, C. (2006). 'The role of science in environmental policy: an examination of the local context'. *Land Use Policy* 23, pp. 161–9.

Johnston, J. L. (2000). *Scotland's Nature in Trust: the National Trust for Scotland and its wildland and crofting management*. Academic Press, London. 266 pp.

Jones, G. (2011). *An investigation into the willingness of recreational hill users to contribute financially towards the costs of upland path repair work*. Unpublished MSc dissertation, University of the Highlands and Islands.

Keating, M. and Stevenson, L. (2006). 'Rural Policy in Scotland after Devolution'. *Regional Studies* 40 (3), pp. 397–408.

Knegtering, E., Hendrickx, L., van der Windt, H. J. and Schoot Uiterkamp, J. M. (2002). 'Effects of Species' Characteristics on Nongovernmental Organizations' Attitudes toward Species Conservation Policy'. *Environment and Behaviour* 34, pp. 378–400.

Lehman, G. (2007). 'The accountability of NGOs in civil society and its public spheres'. *Critical Perspectives on Accounting* 18, pp. 645–69.

Levine, A. (2002). 'Convergence or Convenience? International Conservation NGOs and Development Assistance in Tanzania'. *World Development* 30 (6), pp. 1043–55.

Mc Morran, R., Price, M. F. and McVittie, A. (2006). 'A review of the benefits and opportunities attributed to Scotland's landscapes of wild character'. Scottish Natural Heritage Commissioned Report No. 194 (ROAME No. F04NC18).

Maxwell, J. and Birnie, R. (2005). 'Multi-purpose management in the mountains of Northern Europe – policies and perspectives'. In: Thompson, D. B. A., Price, M. F. and Galbraith, C. A. (eds), *Mountains of Northern Europe: Conservation, Management, People and Nature*. TSO Scotland, Edinburgh, pp. 227–38.

Mitchell, I. (1999). *Isles of the West: A Hebridean Voyage*. Birlinn, Edinburgh. 235 pp.

Mohan, G. (2002). 'The disappointments of civil society: the politics of NGO intervention in northern Ghana'. *Political Geography* 21, pp. 125–54.

Molloy, D., Thomas, S. and Morling, P. (2011). 'RSPB Reserves and Local Economies'. RSPB, Sandy.

National Trust for Scotland (2002). 'Wild Land Policy'. <http://www.nts.org.uk/conserve/downloads/wild_land_policy_2002.pdf> (last accessed 8 January 2013).

National Trust for Scotland (2005). 'Glencoe'. <http://www.glencoe-nts.org.uk> (last accessed 18 January 2013).

National Trust for Scotland (2012a) 'Countryside'. <http://www.nts.org.uk/countryside (last accessed 12 January 2012).

National Trust for Scotland (2012b). 'Mar Lodge Estate Management Plan (2012-2016)'. <http://www.nts.org.uk/Downloads/Properties/management_plan_v18.pdf> (last accessed 8 January 2013).

Natuurmonumenten (2012). 'Natuurmonumenten supporters grow in 2012'. Press release, 7 December 2012. <http://www.natuurmonumenten.nl> (last accessed 19 December 2012)

Pan Parks (2011). 'Best Practice examples of restoring wilderness attributes'. With reference to the third Global Biodiversity Outlook report and Aichi Biodiversity Targets. Pan Parks, Foundation Gyor, Hungary.

Pierce, J. T. (1996). 'The Conservation Challenge in Sustaining Rural Environments'. *Journal of Rural Studies* 12 (3), pp. 215–29.

Price, M. (1994). 'Ecopolitics and Environmental Nongovernmental Organizations in Latin America'. *Geographical Review* 84 (1), pp. 42–58.

Putman, R. J., Duncan, P. and Scott, R. (2008). 'Tree regeneration without fences? An analysis of vegetational trends within the Creag Meagaidh National Nature Reserve 1988–2001'. *Forest Ecology and Management* 206 (1–3), pp. 263–81.

Rissman, A. R., Lozier, L., Comendant, T., Kareiva, P., Kiesecker, J. M., Shaw, M. R. and Merenlender, A. M. (2007). 'Conservation Easements: Biodiversity Protection and Private Use'. *Conservation Biology* 21 (3), pp. 709–18.

Rewilding Europe (2012). 'Executive Summary'. <http://www.rewildingeurope.com/assets/uploads/Downloads/RwE-Factsheet-Juni-2011Def2HR.pdf> (last accessed 20 December 2012).

RSPB (2010a) 'Bringing life back to the bogs; a new beginning for Scotland's majestic flow country'. *Nature's Voice* magazine, RSPB.

RSPB (2010b) 'Futurescapes, space for nature, land for life'. *Nature's Voice* magazine, RSPB.

Scottish Wildlife Trust (2010). 'Official home of the Scottish beaver trial'. SWT, Edinburgh.

Scottish Wildlife Trust (2011). 'Facts and Figures: Conservation on reserves 2010–2011'. SWT, Edinburgh.

Shiel, A., Rayment, M. and Burton, G. (2002). 'RSPB Reserves and local economies'. <http://www.rspb.org.uk/Images/Reserves%20and%20Local%20Economies_tcm9-133069.pdf> (last accessed 8 January 2013).

Simmons, P. J. (1998). 'Learning to Live with NGOs'. *Foreign Policy* 112, pp. 82–96.

SNH (2002). 'Wildness in Scotland's Countryside'. Policy Statement No. 02/03. Scottish Natural Heritage, 2002.

SNH (2009). 'National Trust for Scotland Grant application for Scotland's Mountain Heritage Programme'. <http://www.snh.org.uk/data/boards_and_committees/main_board_papers/0245.pdf> (last accessed 7 January 2009).

SNH (2012a). 'Mapping Scotland's wildness and wild land'. <http://www.snh.gov.uk/protecting-scotlands-nature/looking-after-landscapes/landscape-policy-and-guidance/wild-land/mapping/> (last accessed 8 January 2013).

SNH (2012b). 'Glencoe: disposal of existing visitor centre to NTS and grant aid of interpretation in new visitor centre'. SNH Board Paper, 01/6/12. <http://www.snh.org.uk/data/boards_and_committees/main_board_papers/01612.pdf> (last accessed 8 January 2013)

Scottish Forest Alliance (2012). 'Scottish Forest Alliance Homepage'. <http://www.scottishforestalliance.org.uk> (last accessed 18 January 2012).

Scottish Squirrel Group (2004). *Scottish strategy for red squirrel conservation*. Scottish Squirrel Group. Scottish Natural Heritage, Battleby. 41 pp.

Taylor, P. (2007). 'Wildland benefits; A brief survey of schemes on the wildland network database'. Commissioned report for the Wildland Network for the regional seminar on the economics of wildland, Knepp Estate, April 2007.

VisitScotland (2008). 'The visitor experience 2008'. Harris Interactive.

Warren, C. R. (2009). *Managing Scotland's Environment*. Second Edition. Edinburgh University Press, Edinburgh. 432 pp.

Whitfield, D. P., Fielding, A. H., McLeod, D. R. A. and Haworth, P. F. (2008). 'A conservation framework for golden eagles: implications for their conservation and management in Scotland'. Scottish Natural Heritage Commissioned Report No. 193 (ROAME No. F05AC306).

Wightman, A. (1996). 'Scotland's mountains: an agenda for sustainable development'. Scottish Wildlife and Countryside Link, Perth.

Wightman, A. (2000). *Who owns Scotland?* Third Edition. Canongate, Edinburgh. 237 pp.

Wightman, A. (2010). *The Poor Had No Lawyers: Who owns Scotland and how they got it.* Birlinn, Edinburgh. 320 pp.

Windmill, D., Putman, R. and Maxwell, J. (2011). 'Report for the board of the National Trust for Scotland into the management of deer, woodland and moorland at at Mar Lodge Estate'. Mar Lodge Independent Review Panel, November 2011.

Woodland Trust Scotland (2012). 'In Scotland'. <http://www.woodlandtrust.org.uk/en/about-us/past-present/milestones/Pages/scotland.aspx > (last accessed 15 December 2012).

PART FOUR

Aligning upland estate management with sustainability

CHAPTER EIGHT

A sustainability tool for the owners and managers of upland estates

Jayne Glass

INTRODUCTION

The Land Reform (Scotland) Act opened up a wide range of important debates about land use, initiating more focused thought about how landownership and land management can facilitate practical actions for sustainable land use in Scotland's uplands (Cowell 2006; McCrone et al. 2008; Warren 2009). 'Sustainability' can be interpreted as an essential prerequisite for delivering the wide range of ecosystem services that are linked to upland areas. The concept of sustainability, however, appears not to have permeated fully into the theory and practice of estate management in upland Scotland.

'Sustainability' is not a word commonly encountered in academic or other literature about estate management. Instead, terms such as 'stewardship' or 'responsibility' are more common, and the 'Code of Practice for Responsible Land Management' (2004) of the Scottish Rural Property and Business Association (now Scottish Land and Estates) is a good example. The concept has been dismissed by some as 'empty rhetoric' that means 'anything and everything' (Paterson 2002, 144) and is hard to translate into precise goals (Mather 1999). There is little direct sustainability policy guidance for upland managers (Scottish Executive 2004; Scottish Executive 2005) though, while the Sustainable Estates research was being conducted, the Scottish Government developed a national Land Use Strategy which was laid before Parliament in March 2011 and has 'Principles for Sustainable Land Use' at its core (Scottish Government 2011).[1]

The successful integration of sustainability discourse into estate policy and planning requires a wide range of knowledge about land management and the driving forces that affect landowners' decisions and actions (Berkes 2009; Raymond et al. 2010). Estate owners have the capacity to determine management activities on their land (within the constraints of legislation, policy, management agreements, conservation designations, and so on) and their decisions influence estate environments, economies and communities. In this context, this chapter describes an inclusive and collaborative process to identify sustainability principles and translate them into practice.[2] A sustainability tool comprising twelve indicators was developed over an eighteen-month period, in collaboration with a range of stakeholders interested in upland estate management in Scotland. The tool is designed to enable estate owners and managers to understand how their decisions and actions can positively or negatively affect a range of economic, environmental and social outcomes, and adjust their management practices accordingly. This chapter explains the development and content of the tool in detail.

DEFINING SUSTAINABILITY PRINCIPLES FOR UPLAND ESTATES

The concept of sustainability invites a range of different world views and frames for understanding upland issues, requiring stakeholders to understand more holistically the connections between the environmental, social and economic aspects of the potential of upland areas. Chapter 2 outlined how the Scottish uplands have an important role to play in delivering sustainability principles, listing several policy documents that have made statements to that effect.

Sustainability indicators and the importance of participation

Sustainability indicators are a widely used tool for setting goals and monitoring progress (White et al. 2006), particularly as they can provide evidence for decision-making and reviewing practice. In effect, indicators provide an evidence-based 'goal-set' which helps to define the central tenets of sustainability, guiding policy and projects in any given context (Miller 2001). The Scottish Government, for example, uses fifty National Indicators to monitor progress towards a series of National Outcomes which are linked to the delivery of sustainable development at the national scale (Scottish Government 2012). Indicators can be highly technical: for example, using specific measures to ascertain whether environmental quality is improving or deteriorating, or including socioeconomic and qualitative measures that can help to unpack the complexity of sustainability (Moffatt et al. 2001). Deciding 'what to measure' is a difficult task, and indicator sets are often developed in a 'top-down' manner by academic researchers or policy-makers who do not seek the engagement of stakeholders in the process of indicator selection. As a result, indicators may not be used by land managers and/or local communities, and the criteria that have been used to choose indicators may not be clear (Carruthers and Tinning 2003). Conversely, a 'bottom-up' approach, which engages a range of stakeholders, may not have the capacity to measure sustainability accurately and reliably, though the indicators are more likely to be rooted in an understanding of the local context (Reed et al. 2006). Nonetheless, indicators developed using a 'bottom-up', participatory approach have been shown to be equally or more accurate compared to those developed by experts (Reed et al. 2005; Dougill et al. 2006), and the involvement of a wide range of participants in the development of indicators has been seen as the key to the successful implementation and long-term use of indicators in practice (Bell and Morse 2008).

There are advantages to spending time exploring and understanding people's perceptions of sustainability prior to selecting indicators (Reed et al. 2011). Widely used face-to-face methods, such as workshops and focus groups, offer means to bring a range of people together to do this. These processes are not problem free, however, and difficulties are often encountered when participants represent different interests or expertise, do not have a history of good communication, or suffer from consultation fatigue or doubt about the relevance or credibility of such exercises (Scott et al. 2009). As many people have a stake in how the uplands are managed, the discussion about sustainability is set within a context of diverse public and private preferences. It is therefore necessary to understand the sustainability perceptions of the wide range of upland stakeholders involved in decision-making and practice. The aim of the process described below was to bring together 'expert' and 'local' managerial knowledge in the development of indicators for sustainability on upland estates in order to ensure that the resulting indicators are relevant within local and wider land-use policy contexts, while also encouraging a participatory and deliberative indicator selection process.

Developing a sustainability tool

The participatory development of sustainability indicators for upland estate management took place over four stages, lasting a total of eighteen months between 2008 and 2010, with three to six months between each stage. A Delphi method was used, as explained briefly in Box 8.1 (for a more detailed overview, see Glass et al. 2013 and/or Donohoe and Needham 2008).

Box 8.1 The Delphi method

The Delphi method is a participatory research method used to facilitate a group communication process that addresses areas of limited knowledge surrounding a particular subject (Linstone and Turoff 2002, 3). The Delphi method traditionally involves a 'panel' of participants in an iterative survey to generate consensus or group opinion on a particular topic or policy issue (Donohoe and Needham 2008). The process is anonymous: the members of the panel do not know the identity of other participants (this avoids negative factors, such as the dominance of powerful individuals/groups). Participants are normally asked to complete a written questionnaire by the researcher who collates the responses to the questions and feeds these responses back to the participants for their consideration in a subsequent round, giving each participant the opportunity to adjust their responses accordingly if they so wish. This process is repeated (using multiple rounds) until consensus is reached or questions have been thoroughly discussed. The method effectively enables the researcher to function as a 'process facilitator' who identifies areas of consensus and conflict and feeds back interim results to the panel for further comment and deliberation. In this manner, the members of the panel are challenged to consider the insights and opinions of others, potentially leading them to rethink their individual assumptions.

TABLE 8.1 Individuals participating in the research process (Source: Glass et al. 2013)

Participant	Stakeholder group[a]					Expertise[b]		
	A	B	C	D	E	SUSTY	U/RLU	EM
1	(X)				X	X		X
2	X			(X)			X	X
3	X		(X)		(X)		X	X
4	(X)		X			X	X	
5	(X)	X		(X)			X	X
6	X					X		X
7	(X)		X			X	X	
8				X			X	
9			X				X	X
10		(X)			X	X	X	
11		(X)		(X)	X	X		X
12			(X)		X		X	X
13	(X)			X			X	X
14			X			X	X	
15			X			X	X	
16	(X)			X	(X)	X	X	
17	X	(X)						X
18		X				X	X	X
19	X			(X)				X
Total	5(6)	2(3)	5(2)	3(4)	4(2)	9	14	12

Stakeholder groups – A: practitioners and professionals; B: representative bodies; C: academics and consultants; D: policy-makers and decision-makers; E: non-governmental organisations.
[a] A cross or number in parentheses (X) denotes a previous or other role in that stakeholder group.
[b]*Expertise* – SUSTY: sustainability; U/RLU: upland/rural land use; EM: estate management.

Input from a range of stakeholders

Upland estate management involves a complex range of interest groups and stakeholders, as outlined in Chapters 1 and 2; therefore, participants were chosen to represent a broad range of expertise and experience. Bodies taking an interest in land-use issues in Scotland include: central government and its agencies; the European Commission; international bodies; landowners' organisations (such as Scottish Land and Estates); and environmental/conservation NGOs, such as the RSPB, many of whom collaborate within Scottish Environment LINK. The group of participants (the 'panel') included a mix of nineteen land-management professionals, researchers from different disciplines, policy-makers and members of representative bodies, all with experience and expertise in sustainability, rural/upland land use and/or estate management, preferably on a range of estates. Table 8.1 shows more information about the individuals taking part in the research.

A collective thought process

Figure 8.1 shows the four stages of the research process in more detail (the number of participants who took part in each stage is shown on the right-hand side of the

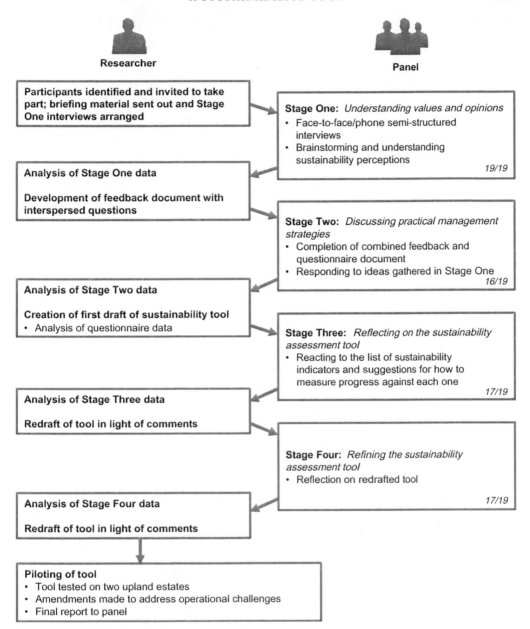

FIGURE 8.1 *Overview of the research process (Source: Glass et al. 2013)*

diagram). In Stage One (Understanding values and opinions), the researcher conducted individual interviews with the participants, asking them to talk about topics and issues that they deemed important in relation to sustainability and estate management, based on their personal and professional experience. Each participant was also asked to suggest specific management strategies that would drive progress towards more sustainable land use. The main themes of these interviews were identified and the results were then presented to the group in Stage Two (Discussing practical management strategies). In this stage, the participants were asked to comment on and expand a list of suggested practical sustainable management strategies that had been identified from the analysis of the results of Stage One.

Using the results of the first two stages, the researcher created the first draft of a sustainability tool which included a list of sustainability indicators and suggestions for how to measure progress against each one. These were presented to the group in Stage Three (Reflecting on the sustainability assessment tool). The participants were then asked to comment on the elements of the first draft; these comments were used to develop a second draft in Stage Four (Refining the sustainability assessment tool). Finally, the framework was amended once more (using the comments made in the final stage) and it was then piloted on two estates in upland Scotland.

PERCEPTIONS AND PRINCIPLES OF SUSTAINABILITY IN ESTATE MANAGEMENT

During the interviews in the first stage, a number of participants expressed negative feelings towards the concept of sustainability. Several participants experienced difficulties with defining its practical meaning and stated a need for more explanation and understanding of the term in the context of estate management. Box 8.2 presents a selection of comments as an illustration.

Box 8.2 Examples of negative feelings/difficulties with the concept of sustainability (direct quotes from participants)

'. . . you'll notice that I've not used the word 'sustainable' at all, and I wouldn't, because I don't understand what it means. It's used all over the place by people who also, I think, do not understand it at all . . .'

'I guess, being slightly cynical . . . the trouble with all this . . . is that you get a buzz word that comes in so, you know, there's acid rain and carbon and climate change and then there's sustainability.'

'There being no definitive definition of sustainability, it's really what you want to make of it – and one person's sustainability is another person's unsustainability. In other words, a particular owner might want to facilitate the ecological element of their estate . . . Conversely, another owner might take a completely different view.'

'You know, my view of sustainable development might be different from a neighbouring estate's view of sustainable development. But, equally, either of them may be valid – it depends on which context you take it in . . . of what spin you put on it.'

'It is always difficult . . . with what is a fluffy concept . . . trying to turn it into something that is tangible and measurable. I think everyone talks grandly about 'sustainable estate management' but then someone like you comes along and says 'define what you mean'. No matter who you ask, everyone is going to come up with a different view.'

Despite these challenges, there was agreement within the group that sustainability should mean a balance between economic, social and environmental aspects of estate management, with the combination of these aspects by estate managers being essential: '. . . we find it to work on the basis of it being a three-legged stool. You need environmental sustainability, economic sustainability and social sustainability. And in the absence of any of those, the stool falls over.' Some participants felt strongly that the estate's natural heritage should take priority as it underpins social and economic activities: '. . . the number of times I've heard: "it's not just about the environment, you know; we have to be sustainable – economically sustainable, socially sustainable", which basically means that it's an excuse to be environmentally unsustainable'. In contrast, others argued that the economic aspects of estate management underpin the environmental and social aspects because financial stability underpins the capacity of the estate owner to manage the land sustainably: '. . . money is a major barrier . . . because an awful lot of estates don't actually make money . . . you've either got to have somebody who'd got funding from elsewhere, or a philanthropist . . . who can continue to fund it'. There was, however, general agreement that focusing heavily on one aspect of management can have negative effects: 'And if you wander around with your blinkers on . . . it doesn't work. It's certainly not sustainable because you're only focusing on management for one particular species or one particular thing. The wheels come off that sort of approach very quickly.' Similarly, 'pigeonholing activities into environmental, social and economic' can limit creative thinking and the ability to 'react to uncertainty'.

Several participants felt that there are emerging opportunities associated with sustainability, and a need to move away from a 'status quo' in order to develop estate businesses and management practices that are more diverse and robust in times of change: 'it is about possibilities . . . about taking off the blinkers and thinking about things in a different way'. It was recognised, however, that estates that are not open to change and to grasping new opportunities are less likely to be sustainable in the long term, and there were particular concerns that cultural issues or mindsets among landowners and managers may hinder progress. Some comments that illustrate this point are included in Box 8.3.

THE TOOL IN DETAIL

Recognising the difficulties inherent in developing a 'one size fits all' definition of sustainability in this context, the participants' comments highlight the need to develop a context-specific, yet flexible, tool for assessing sustainability. This allows the tool to be used on different types of estates with a range of management objectives. Additionally, the group preferred to conceive sustainability as a direction/process rather than as an endpoint or 'destination'. This allows sustainability to be interpreted as providing an approach that can be modified within any context to allow for locally specific solutions and plans. The final sustainability tool comprises four elements that the participants identified and refined during the research process and which help estate owners and managers to understand the important elements of the 'process' of sustainability in estate management (see Figure 8.2):

Box 8.3 Comments about the extent to which cultural attitudes/ mindsets can affect progress towards sustainability

'My initial reaction is to say that . . . the intractable issues here are not the scientific or the intellectual or even the concept of sustainability as much as the clash of cultures.'

'It's a very, very difficult task actually – it's changing a mindset that unfortunately has been there for years.'

'I don't think it's funding . . . sometimes there may be funding constraints around specific concerns. I think it's a mindset definitely. I think if you have the right mindset with the same kind of funding you can make very big differences.'

'The issue of landed business people keeping their cards too close to their chests is a serious constraint. Because they're private businesses with ownership of land, it means that their horizons are "this is my land, what can I do with it?" They don't look for partners.'

However, some participants noted positive changes in this regard:

'I'm sure the old image of a landowner is with a shotgun in one hand and a glass of port in the other, but that is totally untrue. We're all professionally qualified people . . . and to run a landed property you've got to be properly qualified in order to do things right.'

'. . . as the generations pass and the old fuddy-duddies die off, they'll be replaced by more enlightened [people] who are encouraged by fresh initiatives like yours [and] think differently about . . . how they achieve a more fulfilling mix, actually, of objectives'.

- Sustainability principles (5)
- Sustainability 'actions' (12)
- An activity performance spectrum (proactive–active–underactive)
- Identification and understanding of constraining and enabling factors that may affect an estate's performance on each action (factors may include: geographical location; financial resources; funding mechanisms; climate; personal attitudes)

Five sustainability principles identify the core elements of a sustainable approach to upland estate management in Scotland. The twelve sustainability 'actions' set out practical tasks that would assist upland estates in delivering sustainability. These are listed in more detail in Table 8.2. Using a simple performance classification system (proactive–active–underactive), the tool provides an assessment framework for judging how 'proactive' the management approach is in delivering each action. The classification system is explained in more detail in Box 8.4, and wider implications of the 'proactive' class are discussed in Chapter 9.

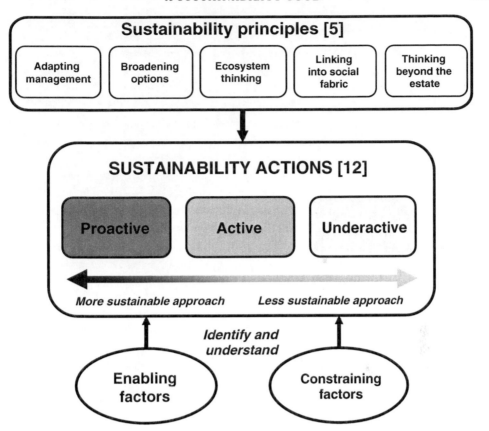

FIGURE 8.2 *Diagram of the sustainability tool developed by the group*

TABLE 8.2 Twelve sustainability actions (indicators) developed by the group

Relates to principle		Sustainability action
Adapting management	1	Long-term, integrated management planning
	2	Integrating monitoring into estate planning and management
Broadening options	3	Adding value to estate business(es), services and experiences
Ecosystem thinking	4	Maintaining, enhancing and expanding natural and semi-natural habitats and species
	5	Maximising carbon storage potential
	6	Maintaining and improving catchments
	7	Maintaining and conserving the estate's cultural heritage
Linking into social fabric	8	Engaging communities in estate decision-making and management
	9	Playing a role in delivering community needs and projects
	10	Facilitating employment and people development opportunities
Thinking beyond the estate	11	Reducing carbon-focused impacts of estate business(es) and other activities
	12	Engaging in planning and delivery beyond the estate scale

Box 8.4 The proactive–active–underactive performance spectrum

The activity performance spectrum allows a judgement to be made regarding whether an estate's management practices are 'proactive' (more sustainable), 'active', or 'underactive' (less sustainable). Throughout the research, the participants regularly highlighted the need to move away from traditional approaches which embrace a 'status quo' in favour of ones that recognise the need for positive change and the management of estates for public, as well as private, benefit. 'Creativity', 'innovation' and 'proactive' attitudes were regularly cited as crucial for a change in the 'mindset' within the estate-management sector, linking back to the comments about cultural attitudes illustrated in Box 8.3. The participants felt that a 'proactive' management approach demonstrates evidence of: going above and beyond conventional good practice; taking a leadership role; and showing courage and/or creativity. In contrast, an 'underactive' approach includes practices that may make an estate more vulnerable to change, or have negative impacts on the local and wider environment or communities. More detail is given in the following table:

Proactive approach *More sustainable*	Active approach	Underactive approach *Less sustainable*
Management goes above and beyond the requirements of statutory/non-statutory guidance	Management complies with statutory/non-statutory guidance	Management fails to comply with statutory/non-statutory guidance
Forward-thinking, creative attitudes ('being the change')	Appropriate responses to adapt to internal and external factors	Lack of response to change when needed (maintaining a status quo for personal preference)
Taking a leadership role at or beyond the estate scale (showing initiative)	Willingness to collaborate with other organisations, partners, etc.	Unwilling to collaborate (closed-door policy)

(Source: Glass et al. 2013)

Participants' comments that illustrate the need for creativity and change:
'the ability to constantly reinvent and change direction to pre-empt external factors'
'the essence of sustainability is the ability to respond to changing circumstances'
'take a positive and proactive approach to opportunities and threats as they appear'
'develop adaptations that can enable managers to maintain viable estates'
'need to make change and not react to it'
'break down their location barrier or turn it into an asset'
'changing approaches and values, not just one business enterprise'

Sustainability Principle 1: Adapting management

Sustainability requires us to consider future generations, recognise the importance of timescales, and be aware of the global context within which management activities sit. As a result, the ability and willingness to adapt to change are central to the concept of sustainable land management, and a joined-up, holistic approach allows a balance of management objectives that can deliver private and public goals. One participant illustrated this well in their interview in the first stage: '. . . really, we're in a changing environment here and therefore anybody should accept that they may have to adapt what they're doing if they want it to be sustainable in the long term. To hope that you can keep everything the same just won't work.' The sustainability actions associated with this principle are shown in more detail in Box 8.5, and a short example is provided.

There is a need for managers who are able to cope with uncertainty and surprise: an adaptive and resilient society requires a critical mass of people who value proactivity (Folke et al. 2005). In the first two stages of the research, participants made comments about the need for estate owners and managers to develop the capacity to adapt to change and develop innovative management strategies in order to become more sustainable. The *'Long-term, integrated management planning'* action highlights the importance of estate managers/owners taking a long-term view, specifically within a formal management plan, which includes clear strategic aims and objectives and which would be reviewed and amended at intervals. At the 'proactive' end of the spectrum, estate managers/owners would seek to develop and review their management plan(s) with local communities and other stakeholders/partners, and the plan should be integrated, combining a range of aspects of estate management, as opposed to creating single documents for some or all of the estate's businesses. Clear, long-term management goals are important so that management objectives are consistent (short-term approaches do not contribute to long-term sustainability – financially, environmentally and otherwise). Similarly, management plans which are reviewed and amended at regular intervals will ensure that management objectives are implemented in an integrated manner and that resources are enhanced and not degraded for current and future generations. Private estate ownership has been billed as having the potential to deliver long-term stewardship: a change in ownership does not necessarily imply a change in management, especially where land agents are involved (Warren and McKee 2011). Frequent changes in ownership can be damaging, however, and the benefits of private ownership in achieving long-term stewardship can be compromised because of the uncertainties of inheritance and owners' motivations (Hunter 2010; Wightman 2010).

The *'Integrating monitoring into estate planning and management'* action highlights the importance of auditing the estate's assets (natural, financial and otherwise) to make informed management decisions. Well-structured monitoring increases awareness and understanding of the assets and facilitates decision-making on the basis of scientific and other evidence. In addition, a connection needs to be made between the results of monitoring and the adaptation of management in the light of new

Box 8.5 Sustainability actions related to the 'Adapting management' principle

PROACTIVE	ACTIVE	UNDERACTIVE
Action 1: Long-term, integrated management planning		
Holistic, integrated, whole-estate management plan	Formal management plan(s) developed	No formal management plan (or ad hoc planning)
For example:	*For example:*	*For example:*
✓ Landscape-scale planning;	✓ Long-term planning;	✓ Lack of plan, or plan out of date;
✓ Strategic, long-term formalised vision for the estate;	✓ Clear actions and objectives;	✓ Short-term or ad hoc decision-making.
✓ Community/external agency involvement in planning;	✓ Broad range of aspects in the plan.	
✓ Consideration of wider impacts of management.		

Action 2: Integrating monitoring into estate management and planning		
Using monitoring results to inform management actions	Carrying out monitoring where required or requested	No resources allocated to monitoring processes
For example:	*For example:*	*For example:*
✓ Taking the initiative to carry out monitoring and use results;	✓ Involvement in statutory monitoring activities;	✓ Lack of involvement in or awareness of statutory monitoring;
✓ Integrating monitoring into risk-management activities;	✓ Responding to and acting on scientific information (e.g. site-condition surveys by SNH);	✓ Lack of aspiration or action to adapt management based on findings.
✓ Involving communities in monitoring;	✓ Contributing to research projects (or allowing research activities on the estate).	
✓ Training staff in monitoring techniques.		

Case study: Action 2 – *PROACTIVE* Integration of monitoring into estate planning

An estate that carries out (and secures funding for) a range of monitoring activities, including: deer numbers (with help from community volunteers); site-condition monitoring; habitat-condition monitoring; species monitoring (golden eagle); and footpath use. The estate also carries out habitat-condition monitoring in a Special Area of Conservation (SAC) which is extra to that required by SNH. There is the aspiration to carry out more surveys and partnerships/joint projects have been pursued with others (e.g. RSPB, SNH) to improve staff knowledge of plant and animal species numbers and populations on the estate.

information (developing strategies that reduce the amount of risk to which activities are exposed). A key element of this action is to gauge the extent to which estate managers monitor change on the estate and integrate the findings into estate planning. Though a high proportion of land managers are interested in the sustainability of natural resources, however, there is a lack of commitment to monitor and a lack of clarity on what standards of habitat quality or impacts are acceptable (Rose 2010, 24). Indeed, one participant commented that monitoring is often anecdotal, rather than formally recorded, recognising that well-structured monitoring is needed to gauge the effects of change.

Sustainability Principle 2: Broadening options

The participants suggested that diversifying (maintaining and widening options) allows flexibility and increases the capacity of estates to adapt to changing circumstances. Linked with this notion are the ideas that financial viability and stability avoid social problems and environmental degradation, and that a narrow range of financial options may lead to the need to draw on capital assets. The sustainability actions associated with this principle are shown in more detail in Box 8.6, and a case study is provided.

'*Adding value to estate products and services*' reflects the need for estate activities to focus on visitors, local communities and other groups to develop unique selling points, services and experiences that could contribute to the financial sustainability of estate businesses. This would ensure that long-term estate-management objectives can be supported financially, reducing pressure on external public/private funding to support estate businesses. This action raises a question with regard to the extent to which estate owners and managers might be willing to add value to their products and services, however, especially when it is difficult to recoup costs or if an addition in value were to lead to an increase in the number of visitors to the estate. A common example is the cost of constructing mountain footpaths which can range from £100 per metre to £200 per metre (for a steep, stone-pitched path) (NTS 2012). While many estates might be able to afford the initial investment, only a small minority can offset these costs with access-related income. As a result, more public funds have become available, normally on a match-funding basis, for access-related expenditure such as signage and footpath repairs (Glass 2010). Private landowners have also been found to have 'little enthusiasm for attracting new clients' as they often value their privacy and personal opportunity to hunt more highly than income (Macmillan and Leitch 2008, 481). Similarly, conservation-minded owners may not wish to increase the economic activity of the estate if this conflicts with conservation objectives.

Sustainability Principle 3: Ecosystem thinking

The 'Ecosystem thinking' principle encapsulates the need for a joined-up, holistic approach, that allows a balance of management objectives to deliver public and private goals. Estate managers are stewards of the natural environment and play an important role in maintaining and enhancing the natural resource base for current and future generations. Similarly, the success of economic and social activities on estates is often underpinned by the productive capacity and/or aesthetic quality of the natural resource base (natural heritage), especially given increasing recognition that

Box 8.6 Sustainability actions related to the 'Broadening options' principle

PROACTIVE	ACTIVE	UNDERACTIVE
Action 3: Adding value to estate business(es), services and experiences		
Developing USPs/brands to develop specialist products and experiences	Adding value to estate activities (financially and otherwise)	Lack of wider marketing of estate business(es) and services
For example:	*For example:*	*For example:*
✓ Developing long-term/ creative income streams;	✓ Reinvesting profit in the estate;	✓ No attempt to attract new investment or income streams;
✓ Diversifying estate business(es);	✓ Recognising the potential of income from visitors;	✓ Supporting non-viable businesses;
✓ Attracting new investment;	✓ Involvement in tourism schemes;	✓ Not recognising the potential of visitor income (or not welcoming visitors).
✓ Promoting the estate;	✓ Estate website.	
✓ Investing in uniqueness.		

Case study: Action 3 – *PROACTIVE* Adding value to estate business(es), services and experiences

An estate that carries out a range of successful activities, including: reinvestment of business profit into the estate; increased profit and viability of estate business(es); added value to tourists' golf and fishing experiences; development projects; marketing promotion activities; business award nominations. Profits received from all aspects of the business are reinvested into the management of the estate's assets and revenue generated from individual businesses is increasing. New business opportunities are also being explored (particularly the potential for selling estate venison). Interaction with potential clients is seen as important, and efforts are made to ensure the customer experience is high quality (branding options have also been explored). A marketing manager has been employed to promote the estate as an area for outdoor activities, at local, national and international scales (in particular, for fishing and golf). Formal business awards and nominations have been received.

the uplands provide a wide range of ecosystem services (as outlined in Chapter 2). The sustainability actions associated with this principle are shown in more detail in Box 8.7, and a short example is provided.

The importance of protecting priority species and habitats is articulated in the *'Maintaining, enhancing and expanding natural and semi-natural habitats and species'* action which requires evidence of favourable conservation status on designated sites, as well as work to restore or expand priority habitats and species. There is a need to protect nationally/internationally important species and habitats, and expand natural capital (according to national, EU and internationally agreed targets). This focus is similar to that of the Scottish Government which had set targets for achieving favourable condition for the special features of designated sites: that 80 per cent of sites

Box 8.7 Sustainability actions related to the 'Ecosystem thinking' principle

PROACTIVE	ACTIVE	UNDERACTIVE

Action 4: Maintaining, enhancing and expanding natural and semi-natural habitats

Working to restore or expand priority habitats and species	Delivering conservation goals in designated areas	Habitat/species degradation, persecution

For example:

For example:	*For example:*	*For example:*
✓ Delivering priority habitat and species protection outside designated areas; ✓ Encouraging and promoting conservation; ✓ Integrating public conservation policies into management actions; ✓ Restoring and creating habitats.	✓ Improving the condition of designated sites and species; ✓ Integrating conservation requirements and potential into management actions.	✓ Lack of action to address habitat degradation affecting priority species and habitats ✓ Persecution of habitats or species.

Action 5: Maximising carbon storage potential

Restoring or expanding carbon stores (peat soils, woodlands etc.)	Controlling degradation of carbon stores (e.g. path management, grazing)	Lack of awareness or management
For example:	*For example:*	*For example:*
✓ Restoring or expanding habitats for carbon storage (e.g. woodlands); ✓ Measuring and/or publicising the carbon storage potential of the estate's assets (raising awareness); ✓ Pre-emptive management of soil erosion (e.g. trampling, overgrazing).	✓ Managing soil erosion (through access management and deer management, for example); ✓ Complying with soil management guidelines.	✓ Active and evident soil erosion with a lack of management to reduce/halt erosion; ✓ Lack of awareness of action on Scottish Forestry Strategy targets (where appropriate).

Action 6: Maintaining and improving catchments

Restoring, expanding or improving riparian zones, wetlands or floodplains	Monitoring and minimising impacts of estate activities on water bodies	Lack of awareness of catchment management or steps to manage pollution or declining water quality
For example:	*For example:*	*For example:*
✓ Enhancing entire riparian zones; ✓ Active role in River Basin Management Planning; ✓ Reducing diffuse pollution through catchment management; ✓ Restoring wetlands/floodplains to reduce flood risk.	✓ Maintaining fresh water resources to good ecological condition.	✓ No awareness of catchment management issues; ✓ Water pollution and/or declining water quality.

Action 7: Maintaining and conserving the estate's cultural heritage

Restoring and/or interpreting cultural heritage assets	Ensuring management activities do not impact upon cultural heritage assets or carrying out conservation where asked	Lack of awareness of cultural heritage or management of sites
For example:	*For example:*	*For example:*
✓ Restoration of cultural heritage sites; ✓ Interpreting cultural heritage for users; ✓ Promoting local heritage (e.g. music, local history, language).	✓ Maintaining cultural heritage assets; ✓ Sympathetic building conversions/changes; ✓ Regular maintenance of listed buildings.	✓ Cultural heritage assets in disrepair; ✓ Lack of interest/action; ✓ Lack of heritage management skills.

Case study: Action 5 – *PROACTIVE* Maximising carbon storage potential

An estate that is working to regenerate and restore areas of natural and semi-natural woodland, as well as renaturalising conifer plantations, restoring mires and ensuring that new planting does not take place on areas of deep peat. Additional activities include a riparian tree-planting project.

would achieve favourable status by 2008 and 95 per cent by 2010. In upland areas, the current levels are 58 per cent 'favourable' or 'unfavourable recovering' (Midgley and Price 2010) which confirms that more effort is required. Increased biological diversity and natural capital increase the potential to draw economic and other benefits from the natural heritage. There can be tensions between managing for one species or habitat rather than for wider biodiversity goals, however, in order to satisfy the personal interest of individuals or organisations (Knegtering et al. 2002). This sentiment is echoed within the tool: a 'proactive' approach to deliver this action involves the delivery of priority species and habitat protection outside designated areas, as well as within them.

Crucially, the link between carbon storage and climate-change mitigation was explored, and participants understood that land can act as a carbon sink, as well as a store, to absorb additional carbon. Carbon storage has moved up the political agenda in recent years, and some participants suggested it may be one of the most important functions of estates in years to come. Storing carbon plays an important role in climate-change mitigation. As Scotland's soils account for 69 per cent of total British carbon storage (Swales 2009), the role of upland estates in managing their soils is clear. Participants also deemed the need to tackle soil erosion as important: once soil is lost or degraded, its ability to deliver primary biological production is very difficult to renew (Haygarth and Ritz 2009). This issue was also raised in 2010, by a variety of lobbying organisations in Scotland, in relation to the growth in the number of hill tracks on upland estates which are built to facilitate easier access to more remote parts of estates, often for stalking purposes and, more recently, for renewable energy developments: a campaign, led by the Mountaineering Council of Scotland, high-lighted how numerous studies in the 1970s and 1980s established that hill tracks have significant detrimental effects on local ecology, water flow, soil chemistry, vegetation patterns, and landscape (Brown 2010).

The *'Maintaining and improving catchments'* action recognises that water man-agement is a key aspect of environmental sustainability in terms of both water quality and the positive role that catchment management can play in flood mitiga-tion, both on and off the estate. Estate-management activities may have an impact on water bodies on the estate (both positively and negatively – for example, improv-ing riparian environments, diffuse pollution) and a joined-up approach to water management (across the catchment) shifts the emphasis from a focus on single issues to thinking about the impacts of management 'downstream'. Water quality is an important issue in Scotland, and further improvements depend largely on the successful management of diffuse pollution from large areas of rural – as well as urban – land (McCracken 2010). Specifically in the uplands, the potential to make the water cycle in river basins operate more naturally through closing drains to rein-state previously drained land, restoring floodplains and floodplain woodlands, and removing canalisation of rivers are all 'practical measures that are known to work' (McCrone et al. 2008, 45). Estate representatives should also be willing to engage in the implementation of hydrological measures, particularly through collabora-tive stakeholder engagement. Therefore, there is potential for 'proactive' practice through playing a role in River Basin/Catchment Management Planning which has developed as a means of implementing the Water Framework Directive (WFD) (Blackstock and Richards 2007).

The *'Maintaining and conserving the estate's cultural heritage'* action includes: the management of cultural heritage (such as archaeological features, buildings, paths); the interpretation of assets (such as signage, ranger services); and recognising the values of landscape and wild land. Cultural heritage assets have intrinsic value and need conserving for current and future generations, and it is possible to draw finan-cial benefits from these assets (for example, through the tourism industry). Each of these aspects recognises the wider public benefits to be gained from 'care for the land-scape and the maintenance of scenic quality' (McCrone et al. 2008, 45), and echoes the fact that resources, such as landscape and the built environment, are components of managed ecosystems. Additionally, there is real scope for these management activities to generate income for the estate through donations and merchandising (for example, the sale of explanatory booklets and donations to support wild land conservation).

Sustainability Principle 4: Linking into social fabric

It was very clear from the outset of the research process that the participants felt that estates play an important role in maintaining rural (and upland) industries and populations, which span landscapes; one estate cannot be considered in isolation. Participants suggested that this 'rural Corporate Social Responsibility' (CSR) role contributes to wider public sustainability goals, in particular by improving quality of life and representation, and livelihood opportunities. The actions associated with this principle are shown in more detail in Box 8.8.

The *'Engaging communities in estate decision-making and management'* and *'Playing a role in delivering community needs and projects'* actions highlight the potential positive role that the landowner/manager can play in collaborating with communities of place and with communities of interest to achieve mutually rewarding goals (this was also the focus of Chapter 5). This can engender a local and wider sense of stewardship and reconnection to natural and cultural heritage, as well as lead to the development of positive relationships between an estate (employees, owner, factor) and its stakeholders/users (by respecting and listening to local knowledge and different types of stakeholders). The participants felt that an improvement in representation was important for engendering a local and wider sense of stewardship and connection to the natural and cultural heritage. This has recently been explicitly emphasised in England's uplands where 'many people are connected economically, socially and culturally to the land and to those who manage the land', and 'this strong, dynamic connection between land and communities is essential in realising the potential of the uplands' (Commission for Rural Communities 2010, 6). Similarly, such practice can maintain and improve the socioeconomic structure of rural communities and improve the quality of life in rural areas. Practical examples of 'proactive' practice include the anticipation of community needs and the initiation and delivery of community-focused projects. The need for affordable housing in rural areas was a key concern of several participants (as was the case in the research reported in Chapter 5), which raises the question of the extent to which the landowner can be involved in delivering more affordable, high-quality rental (or sale) properties on estate land.[3]

The *'Facilitating employment and people development opportunities'* action advocates a more diverse range of stable jobs in rural areas to help maintain population and infrastructure. Estates can facilitate self-sufficient, robust local communities by supporting local trades, products and suppliers, and high-quality jobs with long-term, integrated training opportunities that build local capacity. There is also potential for estate staff to act as 'ambassadors' for the estate and interact with local communities and visitors.

Sustainability Principle 5: Thinking beyond the estate

The weight assigned by participants to the wider impacts of estate management illustrates the need for the management of estates to go beyond the estate unit itself and to connect up with wider landscape-management objectives. This acknowledges the wider regional, national and global context of estate management, and emphasises that collaboration engenders positive attitudes and joined-up thinking to develop new solutions and ideas (promoting accountability and participation). The actions associated with this principle, together with a short example, are shown in more detail in Box 8.9.

Box 8.8 Sustainability actions related to the 'Linking into social fabric' principle

PROACTIVE	ACTIVE	UNDERACTIVE
Action 8: Engaging communities in estate decision-making and management		
Seeking engagement of communities in management	Consulting communities when required AND using results	Lack of engagement with community
For example:	*For example:*	*For example:*
✓ Inviting comment and engagement from communities/users on defining management objectives; ✓ Involving users in practical estate-management tasks (leading); ✓ Inviting external groups to take part in estate activities (learning).	✓ Statutory consultations (using results); ✓ Allowing users to take part in activities on the estate (e.g. volunteer groups); ✓ Allowing school groups to use estate for educational purposes.	✓ Lack of interaction with the community on planning and management activities; ✓ Superficial statutory consultations (not using results or poorly organised process); ✓ Closed-door policy.
Action 9: Playing a role in delivering community needs and projects		
Initiating and/or co-ordinating community projects	Co-operating in community projects	Lack of co-operation in community projects
For example:	*For example:*	*For example:*
✓ Anticipating community needs and taking action; ✓ Providing high-quality rental housing for employees and wider community; ✓ Developing and implementing innovative community projects.	✓ Co-operating in meeting community needs when estate resources are crucial; ✓ Willingness to allow land to be used/leased for community purposes.	✓ Unwillingness to allow use of land/premises when approached by the community; ✓ Lack of communication with community about needs/projects; ✓ Low-quality rental housing stock (or disproportionately high rents).
Action 10: Facilitating employment and people-development opportunities		
Generating significant employment and development opportunities	Supporting a diverse range of jobs	Importing labour when required or lack of investment in people
For example:	*For example:*	*For example:*
✓ Promoting integrated training and developing local skills and confidence; ✓ Providing facilities for training courses, events, other businesses; ✓ Flexible estate staff who work in several activities (ambassadors); ✓ Promoting local supply chains.	✓ Supporting a diverse range of stable jobs; ✓ Training provided for staff; ✓ Using local contractors where available and appropriate.	✓ Lack of formal training for estate staff; ✓ Lack of investment in skills development; ✓ Labour imported when needed, giving little skilled local employment.

Case study: Action 8 – *PROACTIVE* Engaging communities in estate decision-making and management
An estate that embraces potential to involve community members and visitors in practical estate management activities. This includes: conservation work parties; community tree-planting days; offering work experience to school/college students; and consulting the community on renewable energy and housing-development plans. The estate ranger works with local schools to integrate outdoor activities and education into the curriculum, with the aim of reconnecting children to their local environment.

Expanding the CSR role of estates, 'Thinking beyond the Estate' involves '*Reducing carbon-focused impacts of estate business(es) and other activities*'. Estates have a role to play in the wider mitigation of the environmental impacts of human activity – there is scope to consider the potential to move towards low-input systems, and reducing the carbon-focused impacts of estate management can contribute to climate-change mitigation. This action supports Scotland's target to reduce greenhouse gas emissions by 80 per cent by 2050 (compared to 1990), and recognises that land use and changes in land use have great potential either to help or to hinder efforts to meet this, and interim, targets. The European Landowners' Organisation has also recognised the role of estates in this regard, actively encouraging estates to raise the energy efficiency of their buildings and management activities (Dupeux and Rocha 2009).

'Thinking beyond the estate' is also linked to the need for a behaviour change with regard to collaboration and knowledge-sharing within and beyond the estate sector. Participants advocated the importance of '*Engaging in planning and delivery beyond the estate scale*', which challenged the stereotype of estates being unwilling to communicate and share their ideas, highlighting the need for a more communicative management style that promotes accountability and involvement. Joining up thinking and action to involve all types of stakeholders can deliver more sustainable solutions, economies and environments, management conflicts may be avoided or addressed through effective partnerships, and business resilience can be increased through mutually supportive ventures. Collaborative learning also helps to develop the ability to deal flexibly with new situations in order to prepare for uncertainty and surprise (which also reinforces the definition of a 'proactive' manager) (Folke et al. 2005). There was general consensus that the integration of different types of knowledge could help to reduce conflict because this would encourage mutual understanding of estate-management activities. There is also the opportunity for estates to provide on-the-ground learning activities for visitors and the local community in order to raise awareness and promote discussion, as well as develop their own institutional memory, with lessons being learned from previous or other experience(s) so that new ventures can be built on a solid platform and risk managed effectively. Such participation and collaboration are not without problems, however, especially when involvement is not mandatory and non-participation (or superficial participation) can hinder progress. It is with this challenge in mind that participants were emphatic that a 'proactive' approach would include taking a leadership/initiation role in collaborative management/decision-making.

Box 8.9 Sustainability actions related to the 'Thinking beyond the estate' principle

PROACTIVE	ACTIVE	UNDERACTIVE
Action 11: Reducing carbon-focused impacts of estate business(es) and other activities		
Using creative or innovative methods to reduce carbon footprint	Taking steps to increase efficiency of estate activities	Lack of action to reduce carbon-focused impacts
For example:	*For example:*	*For example:*
✓ Investing in sustainable building design; ✓ Investment in sustainable heating systems/micro-renewables; ✓ Other renewable energy projects; ✓ Promoting sustainable choices and lifestyles.	✓ Reducing carbon emissions from estate activities; ✓ Insulating estate buildings (increasing energy efficiency).	✓ Lack of concern or action to reduce environmental impacts of estate businesses or increase energy efficiency.
Action 12: Engaging in planning and delivery beyond the estate scale		
Initiating or co-ordinating collaborative projects, interactions and partnerships	Participating positively in collaborative projects, interactions and partnerships	Lack of collaborative interaction
For example:	*For example:*	*For example:*
✓ Encouraging other partners and/or businesses to enter into partnership or joint business ventures; ✓ Taking a lead role in partnerships or projects, with positive results; ✓ Seeking out opportunities associated with joint working.	✓ Willing to participate in projects and partnerships; ✓ Constructive collaborations, relationships and dialogue.	✓ Lack of participation in projects and partnerships (closed approach); ✓ Irregular or negative commitment to projects or partnerships; ✓ Lack of positive dialogue; ✓ History of unresolved disputes or unwillingness to collaborate.

Case study: Action 12 – *PROACTIVE* Engaging in planning and delivery beyond the estate scale
An estate that contributes to Local Development planning and Community Planning Partnerships acts as lead partner in a local infrastructure development project, engages in lobbying activities and maintains a close working relationship with the local council and enterprise agency.

USING THE TOOL ON THE GROUND

The tool can be used by an external assessor (that is, someone unaffiliated to the estate) or as a self-assessment tool. In both cases, it is imperative that clear, descriptive evidence is used to explain why a particular sustainability class has been assigned, and that decision-making is transparent. The participants favoured the use of the toolkit as an external assessment tool, yet also suggested that there was potential to develop the toolkit further as a self-assessment tool or as a formal benchmarking tool that could be linked to funding applications. The type of estate ownership should not affect the use of the tool: it is designed for use on any estate, whether it is owned privately, publicly, by a community organisation, or by a non-governmental organisation (NGO). It can be regarded as a user-friendly monitoring tool that has been designed to stimulate learning and discussion about how estates manage their resources and how they impact on wider public sustainability goals. Figure 8.3 gives an overview of the process of using the tool.

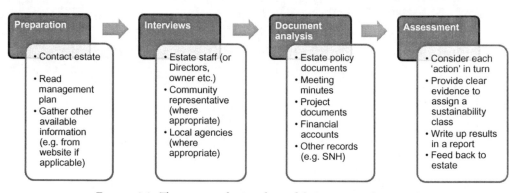

FIGURE 8.3 *The process of using the tool (as an external assessor)*

The following documents are a useful starting point for making judgements about estate-management practice: estate-management plans (such as whole-estate plans, deer-management plans and business plans); estate policy documents and statements; business-strategy documents; planning documents; meeting minutes; financial accounts; newsletters; website (if available). An external assessor can also carry out an in-depth interview with a main representative of the estate. Ideally, this will be a head factor, owner or senior manager with a broad range of responsibilities and good overall knowledge of the estate business(es) and management activities. The tool contains a list of suggested questions in an appendix for exploring each of the sustainability actions in detail. To cross-check the findings of this interview, interviews with other people (for example, other estate staff, local agency officials, community representatives) and appropriate data sources (such as field observations) are also suggested within the tool's appendices. Where the tool is used for self-assessment, these questions can also be used as a guide.

Using the detailed descriptions given for each action, users of the tool assign a sustainability class to each action, based on the evidence gathered. A clear, descriptive narrative should explain why each class has been assigned, and references should be made to other data sources that back up that decision. Results can be entered into the results boxes for each action and 'signposts to sustainability' (areas where improvements or changes could be made) can also be listed. There is also space to reflect on the main enabling and constraining factors that were encountered during the process. In external assessment, the results should then be sent to a main estate representative who should comment on the results, feeding back whether they agree or disagree with the classes that have been assigned. A short note of this meeting should be added in the 'feedback' section. In the case of self-assessment, the feedback section can be used to specify how the results will be used to inform future action planning and management decisions.

Moving forwards

The focus of this research process was on developing an assessment tool and piloting it rather than using it on a range of estates in upland Scotland. The tool in itself provides a framework for sustainable upland estate management, and using the tool more widely would be a valuable exercise that would provide data to compare the extent to which different estates are delivering on the twelve actions. The results of such a study would provide greater insight into the extent to which different estates are embracing a 'proactive' approach and could allow conclusions to be made about the sustainability of different ownership models. The tool in its current format is also a useful learning tool, emphasising the importance of a key person or organisation as instrumental in 'making things happen' (Schultz and Fazey 2009; Scott 2011). By sharing best practice and learning from examples taking place on different estates, the 'proactive' approach may begin to challenge a status quo that was criticised by participants at the outset of the research process. By completing the assessment, an estate's manager is made aware of areas where performance is lacking, and a two-way feedback process, based on the results, allows the development of formal plans to deliver the actions.

As the research process developed and the tool evolved over the four stages shown in Figure 8.1, the participants were increasingly positive about its usefulness and wider application on estates in upland Scotland. At the outset of the process, participants demonstrated much negative feeling towards the concept of sustainability, often dismissing it as too difficult to apply in practice, or so overly complicated that it has lost all meaning 'on the ground'. There was also marked scepticism that it would be possible to develop a generic definition of 'sustainable upland estate management' that would be applicable across all the ownership and management types on Scotland's upland estates. By the end of the process, there was a general feeling of optimism about the progress that had been made, demonstrated by comments such as: 'it seems to be coming together as a good piece of work and a remarkable attempt at defining sustainability in such a varied entity as an estate'; 'it seems a very logical progression from previous work'; and 'I've enjoyed this; I think you have the makings of an extremely valuable tool here'. The participants were very positive about tackling a 'real world' issue and developing a meaningful output; this may help to explain the high levels of motivation throughout the process.

In the responses to the fourth stage, several comments showed how the participants themselves could envisage the practical application of the tool. For example, one commented that 'the toolkit is a very useful document for an estate manager', while another thought that the results should 'be able to inform management usefully'. Others suggested that the research 'might stimulate estates to think about different ways of managing their businesses' and that it was 'a good basis [that] should be worked on in practice on real estates'. Further positive assessments referred to the usability of the classes (proactive–active–underactive) to 'make measurement by non-specialists easy' and as 'a useful approach to framing assessments'.

Interest in the wider application of the toolkit has been shown by a range of private, NGO and community estate owners, as well as from agencies and representative bodies. Scottish Land and Estates, with advice from Scottish Natural Heritage, the Game and Wildlife Conservation Trust, the RSPB and the Cairngorms National Park Authority, is currently developing and piloting a Wildlife Estates Scotland initiative (WES) which will be used by landowners to demonstrate how estates deliver integrated solutions for positive land management and biodiversity. The team developing the WES initiative approached the author for information about the toolkit and subsequently included the proactive–active–underactive spectrum in the pilot version of the WES label.

The use of an iterative, participatory approach ensured that the process of developing the tool was flexible. By asking each participant to reflect on the feedback from each stage, ideas and opinions could evolve over the course of the process, and participants' ideas were repeatedly set within the context of the views of others taking part. This approach facilitated a safe environment for disagreement and debate, prior to developing outputs, and the collaboration of a range of stakeholders brought together, albeit anonymously, a group of people who might not otherwise have discussed sustainability in this context. The implications of the tool and its potential as a framework for sustainable upland management are considered in the next chapter.

Notes

1. There was clearly scope for the findings of this research to feed into the consultation process associated with the development of the Land Use Strategy in 2010. The Sustainable Estates research team submitted a consultation response in December 2010, integrating results from the research process.
2. Initially inspired by the work of Rist et al. (2007) and Ioris et al. (2008).
3. A 'Rural Homes for Rent' grant was introduced as a pilot scheme by the Scottish Government in 2008. This grant allows rural landowners to apply for funding to boost the number of affordable homes available for rent on their land.

References

Bell, S. and Morse, S. (2008). *Sustainability Indicators. Measuring the Immeasurable?* Earthscan, London. 256 pp.

Berkes, F. (2009). 'Evolution of co-management: Role of knowledge generation, bridging organizations and social learning'. *Journal of Environmental Management* 90, pp. 1692–1702.

Blackstock, K. L. and Richards, C. (2007). 'Evaluating stakeholder involvement in river basin planning: a Scottish case study'. *Water Policy* 9, pp. 493–512.

Brown, C. (2010). 'Act Now to Stop the Spread of Hill Tracks'. *Wild Land News* 75, pp. 4–7.

Carruthers, G. and Tinning, G. (2003). 'Where, and how, do monitoring and sustainability indicators fit into environmental management systems?' Special Edition of the *Australian Journal of Experimental Agriculture* 43, pp. 307–23.

Commission for Rural Communities (2010). 'High Ground, High Potential. A Future for England's uplands communities'. Summary report. Commission for Rural Communities, Gloucester.

Cowell, R. (2006). *Towards Sustainable Land Management: A Report for the National Trust.* National Trust, London. 39 pp.

Donohoe, H. M. and Needham, R. D. (2008). 'Moving Best Practice Forward: Delphi Characteristics, Advantages, Potential Problems, and Solutions'. *International Journal of Tourism Research* 11 (5), pp. 415–37.

Dougill, A. J., Fraser, E. D. G., Holden, J., Hubacek, K., Prell, C., Reed, M. S., Stagl, S. and Stringer, L. C. (2006). 'Learning from doing participatory rural research: lessons from the Peak District National Park'. *Journal of Agricultural Economics* 57, pp. 259–75.

Dupeux, D. and Rocha, A. (2009). 'The role of ELO on environmental issues'. Presentation to the European Landowners' Organisation, GA Madrid, 23 November.

Folke, C., Hahn, T., Olsson, P. and Norberg J. (2005). 'Adaptive Governance of Social-Ecological Systems'. *Annual Review of Environmental Resources* 30, pp. 441–73.

Glass, J. H. (2010). 'Access to the hills: five years on'. *The Munro Society Journal* 2, pp. 67–72.

Glass, J. H., Scott, A. S. and Price, M. F. (2013). 'The power of the process: co-producing a sustainability assessment toolkit for upland estate management in Scotland'. *Land Use Policy* 30 (1), pp. 254–65.

Haygarth, P. M. and Ritz, K. (2009). 'The future of soils and land use in the UK: Soil systems for the provision of land-based ecosystem services'. *Land Use Policy* 26S, S187–S197.

Hunter, J. (2010). 'Land for the People: why land reform matters and why we need more of it'. Public lecture, Festival of Politics, Scottish Parliament, 20 August 2010.

Ioris, A. A. R., Hunter, C. and Walker, S. (2008). 'The development and application of water management sustainability indicators in Brazil and Scotland'. *Journal of Environmental Management* 88 (4), pp. 1190–1201.

Knegtering, E., Hendrickx, L., van der Windt, H. J. and Schoot Uiterkamp, J. M. (2002). 'Effects of Species' Characteristics on Nongovernmental Organizations' Attitudes toward Species Conservation Policy'. *Environment and Behaviour* 34, pp. 378–400.

Linstone, H. A. and Turoff, M. (2002). *The Delphi Method: Techniques and Applications.* Addison-Wesley, Reading, MA. <http://www.is.njit.edu/pubs/delphibook/> (last accessed 20 September 2012).

McCracken, D. (2010). 'How effectively is water quality being managed?' In: Skerratt, S., Hall, C., Lamprinopoulou, C., McCracken, D., Midgley, A., Price, M., Renwick, A., Revoredo, C., Thomas, S., Williams, F. and Wreford, A., *Rural Scotland in Focus 2010.* Rural Policy Centre, Scottish Agricultural College, Edinburgh, pp. 60–9.

McCrone, G., Maxwell, J., Crofts, R., Barbour, A., Kelly, B., Linklater, K., Ratter, D., Reid, D. and Slee, B. (2008). 'Committee of Inquiry into the Future of Scotland's Hills and Islands'. Royal Society of Edinburgh, Edinburgh. <http://www.royalsoced.org.uk/cms/files/advice-papers/inquiry/hills/full_report.pdf> (last accessed 30 November 2012).

MacMillan D. C. and Leitch, K. (2008). 'Conservation with a Gun: Understanding Landowner Attitudes to Deer Hunting in the Scottish Highlands'. *Human Ecology* 36, pp. 473–84.

Mather, A. S. (1999). 'The Moral Economy and Political Ecology of Land Ownership'. Paper presented to the Land Reform in Scotland Conference Edinburgh, 29 October 1999.

Midgley, A. and Price, A. (2010). 'What future for upland biodiversity?' In: Skerratt, S., Hall, C., Lamprinopoulou, C., McCracken, D., Midgley, A., Price, M., Renwick, A., Revoredo, C., Thomas, S., Williams, F. and Wreford, A., *Rural Scotland in Focus 2010.* Rural Policy Centre, Scottish Agricultural College, Edinburgh, pp. 80–7.

Miller, G. (2001). 'The development of indicators for sustainable tourism: results of a Delphi survey of tourism researchers'. *Tourism Management* 22, pp. 351–62.

National Trust for Scotland (2012). 'The Footpath Fund – conserving the mountains and wild land you love'. <http://www.nts.org.uk/footpathfund> (last accessed 12 December 2012).

Paterson, A. (2002). *Scotland's Landscape: endangered icon.* Polygon, Edinburgh. 256 pp.

Raymond C. M., Fazey, I., Reed, M. S., Stringer, L. C., Robinson, G. M. and Evely, A. C. (2010). 'Integrating local and scientific knowledge for environmental management: From products to processes'. *Journal of Environmental Management* 91, pp. 1766–77.

Reed, M., Fraser, E. D. G., Morse, S. and Dougill, A. J. (2005). 'Integrating Methods for Developing Sustainability Indicators to Facilitate Learning and Action'. *Ecology and Society* 10 (1), p. 3.

Reed, M. S., Fraser, E. D. G. and Dougill, A. J. (2006). 'An adaptive learning process for developing and applying sustainability indicators with local communities'. *Ecological Economics* 59 (4), pp. 406–18.

Reed M. S, Buenemann, M., Atlhopheng J., Akhtar-Schuster M., Bachmann F., Bastin G., Bigas H., Chanda R., Dougill A. J., Essahli W., Evely A. C., Fleskens L., Geeson N., Glass J. H., Hessel R., Holden J., Ioris A., Kruger B., Liniger H. P., Mphinyane W., Nainggolan D., Perkins J., Raymond C. M., Ritsema C. J., Schwilch G., Sebego R., Seely M., Stringer L. C., Thomas R., Twomlow S., Verzandvoort S. (2011). 'Cross-scale monitoring and assessment of land degradation and sustainable land management: a methodological framework for knowledge management'. *Land Degradation and Development* 22 (2), pp. 261–71.

Rist, S., Chidambaranathan, M., Escobar, C., Wiesmann, U. and Zimmermann, A. (2007). 'Moving from sustainable management to sustainable governance of natural resources: The role of social learning process in rural India, Bolivia and Mali'. *Journal of Rural Studies* 23 (1), pp. 23–37.

Rose, R. D. (2010). 'Sustainable Deer Management: A Case Study Report to the Deer Commission for Scotland'. Deer Commission Scotland, Inverness. <http://www.snh.gov.uk/docs/A438700.pdf> (last accessed 12 December 2012).

Schultz, L. and Fazey, I. (2009). Effective Leadership for Adaptive Management. In: Allan, C. and Stankey, G. H. (eds), *Adaptive Environmental Management: A Practitioner's Guide.* Springer Science and Business Media B.V. 352 pp.

Scott, A. J. (2011). 'Focussing in on focus groups: Effective participative tools or cheap fixes for land use policy?' *Land Use Policy* 28 (4), pp. 684–94.

Scott, A. J., Shorten, J., Owen, R. and Owen, I. G. (2009). 'What kind of countryside do we want: perspectives from Wales, UK'. *Geojournal* DOI: 10.1007/s10708-009-9256-y.

Scottish Executive (2004). 'Planning and Community Involvement in Scotland'. Social Research Unit, Development Department. HMSO, Edinburgh.

Scottish Executive (2005). 'Choosing our future: Scotland's sustainable development strategy'. Scottish Executive, Edinburgh.

Scottish Government (2011). 'Getting the best from our land – A land use strategy for Scotland'. Scottish Government, Edinburgh.

Scottish Government (2012). 'National Indicators'. <http://www.scotland.gov.uk/About/Performance/scotPerforms/indicator> (last accessed 28 November 2012).

SRPBA (2004). 'Code of Practice for Responsible Land Management'. Scottish Rural Property and Business Association, Musselburgh.

Swales, V. (2009). 'The Lie of the Land: Future Challenges for Rural Land Use Policy in Scotland and Possible Responses'. A report commissioned by the RELU programme, January

2009. <http://www.relu.ac.uk/research/Land%20Use%20Consultation/The%20Lie%20of%20the%20Land.pdf> (last accessed 20 December 2012).

Warren, C. R. (2009). *Managing Scotland's Environment.* Second Edition. Edinburgh University Press, Edinburgh. 432 pp.

Warren, C. R. and McKee, A. (2011). 'The Scottish revolution? Evaluating the impacts of post-devolution land reform'. *Scottish Geographical Journal* 127, pp. 17–39.

White, V., McCrum, G., Blackstock, K. L. and Scott, A. (2006). 'Indicators and Sustainable Tourism: Literature Review'. The Macaulay Institute. <http://www.macaulay.ac.uk/ruralsustainability/LiteratureReview.pdf> (last accessed 25 May 2008).

Wightman, A. (2010). *The Poor Had No Lawyers: Who owns Scotland and how they got it.* Birlinn, Edinburgh. 320 pp.

CHAPTER NINE

Lessons for sustainable upland management

**Jayne Glass, Martin F. Price, Alister Scott, Charles Warren,
Robert Mc Morran, Annie McKee and Pippa Wagstaff**

INTRODUCTION

The chapters within this book have investigated the alignment of upland estate management with the concept of sustainability, particularly with regard to the impacts of
the decisions of landowners on upland environments, economies and communities.
Set against the backdrop of increasing rhetoric and policy action in relation to ecosystem services (Chapter 2) and the need for sustainable management of uplands, this
final chapter discusses a number of key issues for further discussion, policy development, and research.

Before the 'Sustainable Estates for the 21st century' project, little academic and
policy attention had been devoted to translating broader sustainability principles
into specific, practical strategies for the management of upland estates. A number
of reasons could be cited, including the great variety of estate-ownership types and
estate-management objectives, and the diversity of the values and opinions held
by the numerous stakeholders. Whatever the reasons are, there is a notable lack of
specific policy instruments and incentives to encourage the owners and managers
of estates to operationalise sustainability. While the idea of a 'sustainable estate' may
seem attractive in theory to those with an interest in the management of Scotland's
uplands, in practice, the terms 'sustainability' and 'sustainable' easily cause confusion,
ambivalence and even animosity as a result of their complex and controversial connotations. Inevitably, stakeholders have different values and management goals that,
in various ways, have an impact on progress towards sustainability, not least because
upland areas provide a great diversity of goods and services to a wide variety of people,
as discussed particularly in Chapters 1 and 2. Nevertheless, the individual studies
reported in full within Chapters 3 to 7, and the development of the sustainability tool
outlined in Chapter 8, point towards five key outcomes that deserve further attention
at various scales, from the owners and managers of individual estates to the Scottish
Government. Each of the following outcomes is discussed in turn in this chapter:

1. 'proactive' estate management (as defined within the tool) is central to sustainable estate management and relies on the presence of a 'locally embedded leader';

2. long-term, integrated management planning is crucial for the development of place-specific interpretations of sustainability, the delivery of ecosystem services, and landscape-scale management;
3. connectivity is a key concept for situating upland estates within the context of wider upland and city regions, concomitant with the need to recognise the wider flows and interactions involving these estates;
4. landowners/managers should give increased priority to a reconstructed role of the estate that incorporates multifunctionality to support socioeconomic development;
5. increasing levels of community engagement in estate-management planning and decision-making create a virtuous circle between economy, community and environment, shaped from the bottom up and leading to more resilient and sustainable outcomes for all parties.

KEY OUTCOMES FROM THE RESEARCH: ALIGNING UPLAND ESTATE MANAGEMENT WITH SUSTAINABILITY

A proactive estate is a sustainable estate

In an address to members of the Scottish Estates Business Group[1] in 2010, the chairman of Scottish Natural Heritage said that 'there is a widely held view that [private] landowners think themselves to be somehow better than everyone else, and this is hugely resented [. . .] it surprises me how few of Scotland's landowners do anything to try and counter this impression' (Thin 2010). The tool described in Chapter 8 acknowledges the importance of a 'proactive approach' to upland estate management which dovetails with this call for estate managers to modify their behaviours, mindsets and values in favour of more integrated management. For example, Bonn et al. (2009a) argue that the notion of maintaining the status quo with regard to landscapes and habitats should be challenged in order to promote the delivery of a wide range of ecosystem services. By conceptualising sustainability as a process, rather than as a goal, it is possible to move beyond the sterile debate about definitions of sustainability to understand the concept through locally derived narratives (Chapters 5 and 6) within which actions can be assessed amid a culture of learning and adaptation.

Through an exploration of the motivations of a sample of private landowners, Chapter 4 demonstrates that there are many examples of proactive practice on privately owned upland estates. The results reveal that, though the options available to many landowners may be limited by economic considerations, the most innovative and entrepreneurial estate owners and managers have been successful in overcoming financial problems, at least in the short term. This research also revealed that absentee landowners, often the most vilified group of landowners, could offer much in terms of providing capital for environmental projects. Chapter 6 also demonstrates the power of community landownership to work as a catalyst for positive change through the impacts on community capacity, cohesion, knowledge and experience relating to resources that are inextricably bound up with their livelihoods. In turn, this helps to address the economic challenges of generating income from the respective asset bases. The importance of a key person or 'locally embedded leader' (Chapter 6), who is instrumental in 'making things happen', has long been recognised, as has the need for managers to be able to cope with uncertainty and surprise (Schultz and Fazey 2009). On both private and community-owned land, the presence of an 'involved' landowner or manager was found to be a significant 'force for good' in helping to develop partnerships and innovative developments/projects.

Such examples of proactive practice show that the owners and managers of estates – whether in private, community or NGO ownership – can adapt to change and develop management strategies in order for the estate to become more sustainable both as a business and in terms of the impacts of management actions on ecosystem services and local communities. This implies a need – and an opportunity – for estates to manage for public, as well as private, benefit(s) and multiple ecosystem services, using novel approaches to play to their geographical, environmental, economic and social strengths. It is this element of 'being the change' that is crucial. While many landowners are sympathetic to such calls, management of this kind costs money, and it may be argued that good practice should be incentivised and/or rewarded in some way by the state. It is here in particular that payments and markets for ecosystem services provide a platform for future policy direction (Rowcroft et al. 2011).

The importance of taking a long-term approach in estate planning

The importance of having long-term land-management goals, as well as the ability to measure progress towards or away from them, is a core component of sustainability, as recognised within the notion of intergenerational equity (Chapter 1). This is perhaps one of the most poorly understood and implemented aspects of sustainability practice, reflecting the corruption of long-term goals by short-term political and economic imperatives. As Vasishth (2008) recognises, to address the realities of sustainable development and environmental change requires more experimentation and risk-taking. Chapters 1 and 2 emphasise the need for a long-term approach to upland estate management, within an ecosystem-services framework. At the heart of such an approach is the development of a coherent estate 'vision' (Chapter 8). The discussion of private ownership in Chapter 3 also highlights the importance of a 'long-term view' in estate management. Frequent changes in ownership can have negative influences on the likelihood of achieving long-term goals, such as environmental protection and socioeconomic development and, while the benefits of private ownership in achieving long-term stewardship have been acknowledged, the uncertainties of inheritance and of owners' attributes present challenges (Hunter 2010). Chapter 4 notes that environmental and social considerations are rarely given a high priority by the private landowners interviewed in this research, posing a challenge to the development and implementation of a truly integrated approach. In practical terms, the success of long-term planning can be constrained by the limited ability of land managers to be reflexive (Low 2002). 'Quick fixes' or 'short-term approaches', often resulting from a lack of financial resources, rarely contribute to long-term sustainability: transformational change occurs through long-term processes that require long-term support, commitment and assessment, as demonstrated on the community-owned sites (Chapter 6). Clear, long-term management objectives that are implemented, reviewed and updated regularly help to ensure that resources are enhanced and not degraded for future generations, and that the provision of ecosystem services is safeguarded. A 'whole estate plan' that integrates management objectives and impacts encourages a balance to deliver public and private goals. Again, the formulation and implementation of such plans can be costly in time and in money, inevitably representing a disincentive when finance is tight, but the potential benefits are considerable. These conclusions derive principally from the research on private estates but apply equally to community- and NGO-owned estates.

A long-term approach is possible only if the financial resources are in place to support it: this can pose a challenge in the upland estate context. Reliance on one or two income streams may make an estate vulnerable to the negative impacts of change, as well as leading to 'corners being cut' in management practice. The precarious nature of finances also means that sympathetic management is by no means guaranteed in the long term (Midgley and Price 2010) however 'enlightened' an owner of any type might ideally wish to be. In the face of declining and uncertain income from 'traditional' activities, such as farming and game management (Chapter 2), estates naturally seek additional income from new enterprises, such as conservation, renewable energy and tourism, particularly as management for game sport (often a loss-making enterprise) remains at the core of most private estates (Chapter 3). Several examples within the preceding chapters have illustrated ways to develop economically viable businesses, and attempts have been made to take advantage of opportunities to diversify private estate activities (Chapters 3 to 5), though, as outlined in Chapter 6, the economic development potential of renewable energy may not always be achievable in practice. Estates owned by environmental NGOs are also likely to become vulnerable owing to their high level of dependence on grants and donated income. Similarly, community landowners, in locations where assets are limited, will need to ensure that they have the capacity to develop their own income streams and become financially self-reliant over the longer term (Chapter 6).

Adding value to estate products and services may allow the development of unique selling points, services and experiences that could enhance the economic resilience of an estate. The Royal Society of Edinburgh's 2008 Inquiry into the Future of Scotland's Hills and Islands recognised the economic importance of tourism and branding, as well as the potential to improve visitor services and outdoor tourism provision (McCrone et al. 2008). Indeed, it is notable that, as reported in Chapter 3, only just over a quarter of the sample of private estates survive without private external financial support – and even with this external 'subsidy', most are only breaking even. There are, however, questions with regard to the extent to which private estate owners and managers may be willing to 'add value' to their products and services, especially when it is difficult to recoup costs: the challenges of investing in improved access have been discussed in Chapter 2. The survey of private landowners also found that the respondents share a set of core values, typically corresponding to 'traditional' management aims and objectives (particularly where sporting activities are concerned), which represent a barrier to innovation. There is, however, an emerging focus on 'sustainable hunting' as a brand/label for good management and conservation to attract more clients, particularly among private landowners across the EU (Dupeux and Rocha 2009), and NGO owners are exploring opportunities associated with 'conservation stalking', venison sales and wildlife photography (Chapter 7).

The importance of connectivity

A key message of the recent United Kingdom National Ecosystem Assessment (UK NEA 2011, 5) is that a sustainable management approach requires 'changes in individual and societal behaviour and adoption of a more integrated, rather than conventional sectoral, approach to ecosystem management': these principles are also embedded in the spatial-planning paradigm (Scott et al. in press). An integrated approach to management recognises that each of the many ecosystem services associated with uplands cannot be successfully managed in isolation. An estate is not an island but a piece of a jigsaw making specific contributions to the wider 'sustainability picture'. Therefore, it is important to identify the specific flows, interactions and dependencies that link each estate with the wider community, environment and economy (Albrechts 2004). There are horizontal (sectoral) and vertical (scale) dependencies that estate plans and strategies must identify and incorporate, with the explicit goal of maximising integration and policy concordance between them. Consequently, there is a need to work across disciplinary, professional and administrative boundaries, mindful of the inequities of power and influence involved (Phelps and Tewdwr-Jones 2000). Thus, action at the landscape scale is a desirable goal to move away from a boundary mentality but there are significant cultural and property-rights barriers to overcome (Chapter 1; Prager et al. 2012).

The landscape scale is arguably the most appropriate at which to understand and assess an estate, integrating the complex interactions between people, place and environment and encouraging 'a more targeted approach to managing the full range of goods and services that uplands offer' (Swanwick 2009, 339). Chapter 1 notes that landscape-scale collaboration normally relies on voluntary co-operation without any statutory footing and, as a result, commitment can often be weak (Prager et al. 2012). Deer Management Groups in upland Scotland (Box 2.6) are a good example. Therefore, there is a need for partnerships with more inclusive representation, explicitly including cross-sector and cross-scalar members from different interest groups, including landowners. Such partnerships will have the mandate to join up policy- and decision-making from the earliest possible stage to find effective ways to manage change and 'identify, promote and deliver visions across administrative and organisational boundaries' (Bonn et al. 2009b, 463; Box 1.5). Moving away from a sectoral approach is not easy but such collaboration can challenge traditional sectoral thinking and enable the sharing of knowledge and expertise, as advocated within the sustainability tool. Cultural and sociopolitical factors, as well as target-led performance criteria, are often the greatest barriers to collaborative management (Koontz and Bodine

2008), and a willingness to share decisions is needed to allow change and to 'think outside the box' (as stipulated in the definition of a proactive manager in Chapter 8). The Scottish Government's Land Use Strategy (LUS: Scottish Government 2011), which aims to facilitate integrated land management, may help to overcome these barriers by enshrining integration as the new norm, rather than the exception, while further incentives towards collaborative approaches may well be built into the next incarnation of the Scotland Rural Development Programme after 2014 (see below).

The multifunctional roles of estates

Estates can play important roles in maintaining the upland economy through interactions with wider communities and business interests. These roles help to support wider public sustainability goals by better aligning community interests and rural planning (Reed et al. 2009), in particular by improving quality of life, livelihood opportunities and the involvement of community members in decision-making (the last of these is considered in more detail below). Given the dominance of private landownership (Chapter 3), concerns continue to be raised with regard to equity and social-justice considerations, access to resources, and rural socioeconomic development. In particular, a major barrier to community sustainability can be the refusal or inability of certain estates to release land for housing and other developments; the former deriving from the landowner, the latter from restrictive rural planning policies (Chapter 5).

Practical examples of 'proactive' practice by estates include the anticipation and identification of community needs within an overall plan and the initiation and delivery of community-focused projects. For example, the encouragement of a diversity of estate enterprises through tenancies had positive implications for local employment on one privately owned estate (Box 5.7, Chapter 5), and the development of affordable housing by community and private landowners goes some way towards tackling the housing provision/affordability challenges outlined in Chapters 2 and 5. In addition, by developing employment opportunities (for example, local renewable energy production), and supporting local trades, markets and suppliers, activities on all types of estates can support more self-sufficient, robust local communities and also provide environmental benefits.

There has been a slow but steady decline in the number and range of services provided locally in rural areas (McCrone et al. 2008; Commission for Rural Communities 2010), highlighting the need for creative, multifunctional local solutions that can reduce the distance that people need to travel to obtain food and other goods and services (Slee 2009). Estates are able to support local economies and employment opportunities, with specific examples of good practice including the generation of significant employment in the local economy and the promotion of integrated training and skills development. This is important because the existence of a diverse range of stable jobs in rural areas can help to maintain population and infrastructure (Chapter 5). The examples from community landownership in Chapter 6 show how such owners are prioritising investment in infrastructural and housing initiatives, supporting local entrepreneurial activity, generating a diverse range of skilled jobs, and reconstructing relationships between communities and their environment. Crucially, the shift away from the idea of a 'preserved wilderness' towards one of the 'working wild' (Chapter 6) is encouraging community landowners to investigate proactive and innovative ways of engaging (and 'consuming') the environment, and to recognise their potential contribution to local development. These examples offer an important signpost for private and NGO-owned estates to strengthen their multifunctional roles within the rural economy.

The virtuous circle of community engagement and collaboration

Several of the chapters have pointed towards the need for more collaboration and stakeholder involvement in the management of upland estates, and the sustainability tool emphasises the importance of 'engaging communities in estate decision-making and management' as one of the twelve sustainability actions. Chapter 5 shows that positive relationships and mutually acceptable outcomes tend to occur when an estate moves away from a 'paternalistic' role and shows genuine interest in community matters, adopting the role of 'partner' and aiming to interact with and involve community members on equal terms.[2] When owners or managers of private and NGO estates increase their engagement and collaboration with local communities, leading to the delivery of mutually agreed practices/projects, this helps to 'reconnect' local and other people to the land on these estates, as well as allowing dialogue and power sharing that ensure estate management practices are responsive to a range of needs. Community engagement represents the embodiment of community landownership; nevertheless, examples of conflict resulting from insufficient engagement are evident even on community landholdings (Chapter 6), highlighting the critical importance of continual transparent and empowering engagement across all forms of rural governance. All landowners could usefully consider the potential mutual benefit of a range of approaches: community consultation 'surgeries'; establishing a community–estate liaison group; estate open days; joint action planning; or an estate newsletter/website (Figure 5.2). Such engagement activities have the potential to deliver value for money (and effort) in the long term by developing social capital and wider community understanding and support for estate activities. The examples of extensive activities carried out by rangers on community-owned estates also show how rangers can function as 'ambassadors' for the community landowning body and play a key role in providing local people and visitors with opportunities to experience their local environment (Chapter 6).

Community involvement in land-use decisions is increasingly being advocated in national assessments and being incorporated in public policy. Both the Programme of Work on Mountain Biodiversity of the Convention on Biological Diversity (CBD 2004) and the chapter on mountain ecosystems in the Millennium Ecosystem Assessment (Körner et al. 2005) state the importance of involving local communities and all relevant stakeholders in the formulation and implementation of visions for upland management, as well as using a range of knowledge(s) to inform policy responses. The key messages of the 2011 United Kingdom National Ecosystem

Assessment also emphasise the importance of dialogue and collaboration (UK NEA 2011). One of the outcomes envisaged in Scotland's National Performance Framework is the growth of 'strong, resilient and supportive communities, where people take responsibility for their own actions and how they affect others' (Scottish Government 2013). The LUS also places emphasis on the need for all communities (urban and rural) to be better connected to the land, with more people enjoying the land and positively influencing land use (Scottish Government 2011). While noting that estates engage with their communities to varying degrees, the LUS welcomes the progressive approaches increasingly adopted by many estates. In addition, it recognises that there should be 'opportunities for all communities to find out about how land is used, to understand related issues, to have a voice in debates, and if appropriate to get involved in managing the land themselves' (Scottish Government 2011, 25).

Promoting community empowerment to stimulate and harness the energy of local people is also a prevalent rural policy theme enshrined in the Scottish Community Empowerment Action Plan 2009, the Land Reform (Scotland) Act 2003 (as discussed in Chapter 3), and the forthcoming Community Empowerment and Renewal Bill. This will include action to build community capacity, recognising the particular needs of communities facing multiple social and economic challenges. The concept of community resilience (Magis 2010) recognises the need for the intentional development of personal and collective capacity so that communities can engage with, respond to, and influence change.

Crucially, communities may need to develop their own narratives and trajectories in order to sustain and shape their futures (Mackenzie 2012). Given the predominance of private ownership in Scotland's uplands, this may often best be done in co-operation with local estates. Given the lessons from the community landholdings in Chapter 6, however, in some cases, the empowerment of communities through the purchase of estates may represent their best options for the future. In general, power imbalances remain a fundamental constraint on the successful development and implementation of partnership working between private estates and their estate communities. For example, the recent 'Uplands Solutions' report (Scotland's Moorland Forum 2011) highlighted a lack of integration between the aspirations of private landowners, land managers and community members as a barrier to progress, and showed, for one upland community, that no links exist between the wide range of community groups and those who manage the surrounding land. Representatives of private estates can play key roles in furthering the development of community resilience, and their effective management of interactions with estate communities and wider communities of interest can have direct and indirect impacts on the sustainability of the outcomes of estate management activities more generally. For example (as identified in the sustainability tool and Chapter 5), involving local communities in private estate management planning can bring mutual benefits; the development of creative business opportunities on the estate can add value to the estate business while also developing new employment for community members.

Confronting these problems explicitly, the following key actions are proposed to engender mutual support, respect, engagement and trust among estate representatives, local communities and other stakeholders. They are suggested for estate owners and managers as aids to decision-making processes that explore mutually beneficial projects, processes and outcomes, rather than as a 'one size fits all' toolkit. These points are applicable to all types of estates, particularly those owned privately or by NGOs.

First, all partners need to take a 'proactive' approach, set within a long-term and inclusive vision formulated in conjunction with local communities and wider stakeholders. To enhance community capacity and overcome contrasting perspectives, trusting relationships are a *sine qua non*. It is critical for community organisations to be open to including representatives of private and NGO-owned estates so that they can develop new relationships, and to explore the benefits of working more closely with agencies and other stakeholders. All partners need to adopt professional attitudes and contribute equivalent time and resources where possible. Ultimately, relationship-building should focus on the development of mutual trust and respect for the differing roles and interests of the different groups and individuals, as well as the development of open and transparent engagement processes, as discussed in Chapter 5 and shown in Figure 5.2.

Interactive (rather than 'top-down') communication is imperative, and the shared development of plans for new developments and initiatives should start as early as possible, and be meaningful and transparent. On private and NGO-owned estates, communities are unlikely to expect involvement in day-to-day estate management activities, though this is the case on some estates. The creation of a Liaison Group on the Dundreggan Estate (Trees for Life; Chapter 7) provides a good example. Representatives of private and NGO-owned estates also need to be visible and approachable; this can be a particular challenge where land is owned by 'secretive offshore companies and anonymous trusts' whose representatives are hard to locate (Wightman 2010, 301). The research on the motivations of private landowners (Chapter 4) indicates that, overall, community-based entrepreneurship and diversified activities increase where the landowner is resident on the estate and plays an active part in the day-to-day management of estate business(es). An 'open-door policy' and good visibility of the estate's representatives are crucial for developing a positive relationship between the estate and the estate community (Chapter 5). Similarly, estates that have proactively developed relationships with wider partners, such as community development trusts, public agencies and NGOs, are better able to access wider support, knowledge and resources. This may pose a challenge, however, on estates where the landowners do not consider themselves to be part of a local community (Chapter 4).

Finally, the research has pointed in more detail towards the need to explore suitable models for rural leadership. Partnerships can increase the capacity of communities and estates to collaborate in local decision-making and develop increased confidence, and effective leadership can release volunteer energy and increase the availability of local social capital. By sharing responsibility among estate, community and other partners, a sense of reliance on the estate can be replaced with more productive, reciprocal and mutually beneficial action.

LOOKING AHEAD: IMPLICATIONS FOR POLICY, PRACTICE AND FUTURE RESEARCH

Large, privately owned upland estates are likely to remain a key feature of Scotland's physical and socioeconomic landscape for the foreseeable future. Nevertheless, there is now an expanding 'middle ground' in which new models of ownership and management are being explored. As a result, the future pattern of landownership and land management in Scotland is likely to be more diverse, flexible and experimental than for many centuries, suggesting that the Land Reform (Scotland) Act 2003 (LRSA) may ultimately help to deliver the sustainability goals that were its avowed objective (Warren and McKee 2011).

Community buyouts are set to continue in Scotland, with the Scottish Land Fund re-established in June 2012 and Scottish ministers appointing a Scottish Land Reform Review Group (LRRG) in the following month. The results of the private landowner survey and interviews with expert commentators (Chapter 3) suggest that the passage of the LRSA has not yet significantly influenced the motivations behind private purchase and retention of upland estates, and that cost remains an important factor in management decision-making. Estate owners are also bound by regulations within the Wildlife and Natural Environment (Scotland) Act 2011 (WANE Act) that collates and updates much game law and regulatory requirements for upland management, and is of particular relevance to private sporting estates in the uplands. A key contested element of the Act was the provision of 'vicarious liability' on the part of private landowners in cases of wildlife crime on their land committed by estate employees unless the landowner shows due diligence in preventing such crimes (Scottish Land and Estates 2011a; Scottish Government 2012). It may also be argued, however, that the continuing political power of private landowners prevents other radical changes with potentially negative impacts on traditional sporting practices, including a proposed ban on snaring (Edwards 2011; Hogg 2011).

As noted frequently in this book, public policy in Scotland increasingly focuses on the need for land management to deliver multiple, integrated benefits (both private and public), which adds new dimensions to the scope and impacts of estate management decisions. The LUS is built on the premise that land use should provide multiple benefits across various spatial scales while also protecting and enhancing ecosystem services. Private landowners maintain that such an integrated approach is not new to private land management, and nor is private inward investment that indirectly contributes to the wider public good (see, for example, Warren 2009; Scottish Land and Estates 2011b). At present, uptake and implementation of the principles of the LUS are voluntary, and private landowners can only be 'strongly encouraged' to take steps towards its implementation (see Box 2.3). Nonetheless, payments for ecosystem services (PES) are widely under consideration, and the programme that follows the current Scotland Rural Development Programme (SRDP) may include a funding stream to incentivise the provision of public goods, permitting the targeting of land managers and locations that provide required ecosystem services. Even though appropriately designed government policy has the potential to influence some landowners' decision-making processes through financial incentives, many of the private owners studied in Chapter 4 prefer to retain their freedom of manoeuvre by not relying on subsidies and, among these owners, most environmental activities are being carried out by those who are financially independent. Undoubtedly, the WANE Act and the LUS have reanimated long-standing debates regarding the private provision of public goods, the public rewards and incentives (if any) that should be offered, and the potential impacts on the institution of private property.

At the same time, given the challenges facing private landownership – including the perceived threats to private property posed by the LRSA, and negative public perceptions of private landowners, as outlined in Chapter 3 – the establishment of community engagement and collaborative processes is likely to be critical for estates to align with 'instrumental imperatives' (Stirling 2006). As outlined above, the involvement of communities and other stakeholders in estate management is an 'important way to sustain or restore public credibility' (Stirling 2006, 96), and the research described in this book has highlighted other potential benefits of the instigation and careful implementation of community engagement processes for all types of landowners. These benefits include: better decision-making owing to the inclusion of a wider range of perspectives and expertise, including local knowledge (Reed 2008; Irvine et al. 2009); increasing the potential for innovation (Brandenburg et al. 1995, in Carr and Halvorsen 2001): and greater support for land-management practices and estate developments through increased public understanding. There is an opportunity for additional government support to maximise the contribution of private estate/community partnerships to sustainability, and this may be incorporated in the forthcoming review of land-reform policy. Such steps may require support to overcome barriers (including the uncertainty of legislative change and taxation levels) to tenancies and joint enterprises by creating flexible ownership mechanisms that encourage shared management and equity (Chapter 5). As participatory governance is considered to help deliver the 'public good' (Carr and Halvorsen 2001), improved partnership processes or joint ownership mechanisms may also support land managers in complying with, and being supported by, the current legislative framework, including the LUS, as well as the Community-led Local Development approach included in the European Commission's 2014–20 Strategic Framework (European Commission 2011).

The research has also highlighted the important and increasing role of the planning system in helping to enable new visions of sustainability to be translated into actions on the ground. As estates diversify, their new ventures bring them into direct contact with the planning system, demanding integration with relevant development-plan policies. The planning system has a legacy of protecting the countryside for its own sake, and wider rural development has struggled to assert its own identity, given the predilection towards agriculture and forestry activities. Set within a reconstruction of rurality and the new paradigm of spatial planning (as discussed in Chapter 1), which stress the need for proactivity, integration, inclusion, long termism and connectivity (Tewdwr-Jones et al. 2010; Scott et al. in press) – the very principles emerging from this research – it is important for planners to work together with estate managers in helping to join up policy and decisions regarding plans for, and the management of, estates.

Conclusions

A sustainable future for Scotland's uplands relies on creating support for 'shared visions and the adoption of an integrated approach' (Bonn et al. 2009a: 475), fully involving the key stakeholders and their plans, policies and subsequent decision-making. Who owns land and how land is managed are distinct, but closely intertwined, dimensions of the challenge of aligning upland management with sustainability: there is no doubt that the ways in which land is owned and used have far-reaching implications, not only in socioeconomic and environmental terms, but also culturally, politically and even spiritually (McIntosh 2001; Warren and McKee 2011). Landownership will always be a key influence on the sustainable management of the Scottish uplands. Unavoidably, land reform and land management involve power relationships: the ownership of land in Scotland gives a measure of power (political, social and economic) to the owner or owners (Chapters 3 to 5; Pillai 2012), and community buyouts have initiated an ongoing renegotiation and reconstruction of power relations which, like yeast working through dough, could ultimately have a transformative impact (Chapter 6; Satsangi 2009). Land-reform legislation has undoubtedly accelerated the emergence of several models of ownership in the Scottish uplands, even if the formal legal procedures have been used quite rarely. 'Social ownership' by communities, in some cases in co-operation with NGOs, will continue and is helping to create increased diversity in landownership and land management. No single model is 'the answer'; as has long been recognised, a plurality of ownership and management types is likely to offer the surest route to a genuinely sustainable future because 'only a pluralistic solution can engage the interest and benefit from the talents of the maximum number of people' (Raven 1999, 150). The timescale over which the success of newer types of ownership should be judged is decades or even centuries; the sustainability tool, developed as part of this project, could assist in assessing the extent to which different estates deliver the twelve actions comprising the tool. More widely, estates and their decision-making processes do not sit in isolation from other institutional structures; it is important that regulatory processes support and enable future reconstructions and renegotiations of estates in order to achieve their potential. Chapter 6 provides important indications of the importance of the planning system in this context, and suggests that planning reform is overdue.

With growing recognition of the key roles of the uplands in providing ecosystem services, it is important to recognise how they 'belong' to a wider community than just those who live in and/or own land in these distinctive and highly valued landscapes. In past times, landowners were free to exercise 'exclusive dominion' over their domains but their powers are now significantly and increasingly constrained within contemporary land-use policy and regulation (Chapter 2). This permits a reconceptualisation of the potential of Scotland's upland estates to deliver a range of integrated benefits for conservation, local communities and the wider public good while, at the same time, generating income to safeguard long-term financial sustainability. The findings of the research reported in this book emphasise the importance of integrating various objectives of estate management and of involving key players, including local communities, in making decisions concerning the management of estates. For private and NGO-owned estates, partnerships between estates and communities can deliver a wider range of social, environmental and economic benefits and strengthen the sustainability of estate management. In this context, the model of joint ownership and management by partnerships involving local people and national conservation organisations, explored in Scotland as well as in other countries around the world, may be of particular relevance even though this model is neither a panacea nor universally applicable. Considerable lessons can be learned from the case studies of community landholdings (Chapter 6) with respect to meaningful and empowering engagement and partnership working. Collaborative initiatives between all types of estates (both individually and across landscapes), communities and other partners, such as government agencies and NGOs, offer considerable potential to deliver rural services, such as affordable housing, enable the development of community and private enterprises, and empower rural communities.

We recognise that these conclusions derive from detailed research on a relatively small number of upland estates, owned by private individuals, community organisations, and NGOs, as well as our larger sample of private estates, and that the research considered neither estates owned by government agencies nor the key roles of land agents in managing many estates across the uplands of Scotland. Furthermore, while we recognise the ideological and highly political questions regarding the 'sustainability' of estates in Scotland, as advocated by many land-reform activists, the project sought to maintain an objective and exploratory research focus, on the basis that providing greater knowledge and understanding of how best to maximise the opportunities and benefits of the current system – including new models of partnership being developed by some private estates and as represented by community- and NGO-owned estates – provides a secure foundation for constructive discussion and action. Nevertheless, the scope for further research is considerable. In conclusion, we believe that our findings have relevance not only for the uplands but also more widely in rural Scotland and beyond – not least in highlighting the opportunities for land-owners of all kinds to forge mutually beneficial and sustainable partnerships with the people with whom they share the privilege of living and working amid the challenging yet special landscapes of Scotland's uplands.

NOTES

1. The Scottish Estates Business Group (SEBG) is no longer a distinct organisation; it is now part of Scottish Land and Estates.
2. Further discussion of community engagement in private estate management can be found in a short, practical booklet entitled *Working Together for Sustainable Estate Communities: Exploring the Potential of Collaborative Initiatives between Private Estates and Communities*

(Glass et al. 2012). A draft of the booklet was presented at three workshops in upland Scotland in October/November 2011; following feedback on its content, the final version is available from <www.perth.uhi.ac.uk/sustainable-estates>.

References

Albrechts, L. (2004). 'Strategic (spatial) planning re-examined'. *Environment and Planning B: Planning and Design* 31, pp. 743–58.

Bonn, A., Allott, T., Hubacek, K. and Stewart, J. (2009a). 'Conclusions: Managing change in the uplands – challenges in shaping the futures'. In: Bonn, A., Allott, T., Hubacek, K. and Stewart, J. (eds), *Drivers of Environmental Change in Uplands*. Routledge, London and New York, pp. 475–94.

Bonn, A., Rebane, M. and Reid, C. (2009b). 'Ecosystem services: a new rationale for conservation of upland environments'. In: Bonn, A., Allott, T., Hubacek, K. and Stewart, J. (eds), *Drivers of Environmental Change in Uplands*. Routledge, London and New York, pp. 448–74.

Carr, D. S. and Halvorsen, K. (2001). 'An Evaluation of Three Democratic, Community-Based Approaches to Citizen Participation: Surveys, Conversations With Community Groups, and Community Dinners'. *Society and Natural Resources* 14, pp. 107–26.

Commission for Rural Communities (2010). 'High Ground, High Potential. A Future for England's uplands communities'. Summary report. Commission for Rural Communities, Gloucester.

Convention on Biological Diversity (2004). 'COP 7 Decision VII/27: Mountain Biodiversity'. <http://www.cbd.int/decision/cop/?id=7764> (last accessed 20 December 2012).

Dupeux, D. and Rocha, A. (2009). 'The role of ELO on environmental issues'. Presentation to the European Landowners' Organisation, GA Madrid, 23 November 2009.

Edwards, T. (2011). 'Wildlife and Natural Environment (Scotland) Bill: Stage 3'. SPICe Briefing, The Scottish Parliament, 22 February 2011.

European Commission (2011). 'Community-led local development. <http://ec.europa.eu/regional_policy/conferences/od2012/doc/community_en.pdf> (last accessed 20 December 2012).

Glass, J. H., McKee, A. and Mc Morran, R. (2012). 'Working Together for Sustainable Estate Communities: exploring the potential of collaborative initiatives between privately-owned estates, communities and other partners'. Centre for Mountain Studies, Perth College, University of the Highlands and Islands.

Hogg, A. (2011). 'The Wildlife and Natural Environment Bill – A Summary of Changes'. The Scottish Gamekeepers Association. <http://www.scottishgamekeepers.co.uk/content/wildlife-and-natural-environment-bill-summary-changes> (last accessed 20 December 2012).

Hunter, J. (2010). 'Land for the People: why land reform matters and why we need more of it'. Public lecture, Festival of Politics, Scottish Parliament, 20 August 2010.

Irvine, R. J., Fiorini, S., McLeod, J., Turner, A., van der Wal, R., Armstrong, H., Yearley, S., and White, P. C. L. (2009). 'Can managers inform models? Integrating local knowledge into models of red deer habitat use'. *Journal of Applied Ecology* 46, pp. 344–52.

Koontz, T. M. and Bodine, J. (2008). 'Implementing ecosystem management in public agencies: lessons from the US Bureau of Land Management and the Forest Service'. *Conservation Biology* 22, pp. 60–9.

Körner, C., Ohsawa, M., Spehn, E., Berge, E., Bugmann, H., Groombridge, B., Hamilton, L., Hofer, T., Ives, J., Jodha, N., Messerli, B., Pratt, J., Price, M., Reasoner, M., Rodgers, A., Thonell, J. and Yoshino, M. (2005). 'Mountain systems'. In: Hassan, R., Scholes, R. and Ash, N. (eds) *Ecosystems and Human Well-being: Current State and Trends*, Volume 1. Millennium Ecosystem Assessment, Island Press, Washington, DC, pp. 681–716.

Low, N. (2002). 'Ecosocialisation and environmental planning: a Polanyian approach'. *Environment and Planning A* 34 (1), pp. 43–60.

McCrone, G., Maxwell, J., Crofts, R., Barbour, A., Kelly, B., Linklater, K., Ratter, D., Reid, D. and Slee, B. (2008). 'Committee of Inquiry into the Future of Scotland's Hills and Islands'. Royal Society of Edinburgh, Edinburgh. <http://www.royalsoced.org.uk/cms/files/advice-papers/inquiry/hills/full_report.pdf> (last accessed 30 November 2012).

McIntosh, A. (2001). *Soil and Soul: people versus corporate power.* Aurum Press, London. 384 pp.

Mackenzie, A. F. D. (2012). *Places of Possibility: Property, Nature and Community Land Ownership.* Wiley–Blackwell, London. 270 pp.

Magis, K. (2010). 'Community Resilience: An Indicator of Social Sustainability'. *Society & Natural Resources* 23 (5), pp. 401–16.

Midgley, A. and Price, A. (2010). 'What future for upland biodiversity?' In: Skerratt, S., Hall, C., Lamprinopoulou, C., McCracken, D., Midgley, A., Price, M., Renwick, A., Revoredo, C., Thomas, S., Williams, F. and Wreford, A., *Rural Scotland in Focus 2010.* Rural Policy Centre, Scottish Agricultural College, Edinburgh, pp. 80–7.

Pillai, A. (2012). 'Land Law'. In: Mulhern, M. (ed.), *Scottish Life and Society: A compendium of Scottish Ethnography*: Volume 13. Institutions of Scotland – The Law. University of Edinburgh, Edinburgh, 367–88.

Prager, K., Reed, M. S. and Scott, A. J. (2012). 'Viewpoint: Encouraging collaboration for the provision of ecosystem services at a landscape scale – Rethinking agri-environmental payments'. *Land Use Policy* 29, 244–9.

Raven, H. (1999). 'Land reform'. In: McDowell, E. and McCormick, J. (eds), *Environment Scotland: Prospects for Sustainability.* Ashgate, Aldershot, pp. 139–53.

Reed, M. S. (2008). 'Stakeholder participation for environmental management: A literature review'. *Biological Conservation* 141, pp. 2417–31.

Reed M. S., Bonn A., Slee W., Beharry-Borg N., Birch J., Brown I., Burt T. P., Chapman D., Chapman P. J., Clay G., Cornell S. J., Fraser E. D. G., Glass J. H., Holden J., Hodgson J. A., Hubacek K., Irvine B., Jin N., Kirkby M. J., Kunin W. E., Moore O., Moseley D., Prell C., Price M. F., Quinn C., Redpath S., Reid C., Stagl S., Stringer L. C., Termansen M., Thorp S., Towers W. and Worrall F. (2009). 'The future of the uplands'. *Land Use Policy* 26S, S204–S216.

Rowcroft, P., Smith, S., Clarke, L., Thomson, K. and Reed, M. (2011). 'Barriers and Opportunities to the Use of Payments for Ecosystem Services'. Final Report to Defra. URS Scott Wilson, The James Hutton Institute, University of Aberdeen.

Satsangi, M. (2009). 'Community land ownership, housing and sustainable rural communities'. *Planning Practice and Research* 24 (2), pp. 251–62.

Schultz, L. and Fazey, I. (2009). 'Effective Leadership for Adaptive Management'. In: Allan, C. and Stankey, G. H. (eds), *Adaptive Environmental Management: A Practitioner's Guide.* Springer Science and Business Media B.V. 352 pp.

Scotland's Moorland Forum (2011). 'The Upland Solutions Project Final Report'. Scotland's Moorland Forum, Dumfries.

Scott, A. J., Carter, C. E., Larkham, P., Reed, M., Morton, N., Waters, R., Adams, D., Collier, D., Crean, C., Curzon, R., Forster, R., Gibbs, P., Grayson, N., Hardman, M., Hearle, A., Jarvis, D., Kennet, M., Leach, K., Middleton, M., Schiessel, N., Stonyer, B. and Coles, R. (in press). 'Disintegrated Development at the Rural Urban Fringe: Re-connecting spatial planning theory and practice'. *Progress in Planning.*

Scottish Government (2011). 'Getting the best from our land: A Land Use Strategy for Scotland'. Scottish Government, Edinburgh.

Scottish Government (2012). 'Protecting our wildlife'. <http://www.scotland.gov.uk/News/Releases/2012/01/22123346> (last accessed 20 December 2012).

Scottish Government (2013). 'National Outcomes – Communities'. <http://www.scotland.gov.

uk/About/Performance/scotPerforms/outcome/communities> (last accessed 20 December 2012).

Scottish Land and Estates (2011a). 'Landowners urged to "answer the call" of wildlife'. Scottish Land and Estates Press Release, Ramsay Smith, Media House, 1 July 2011.

Scottish Land and Estates (2011b). 'Wildlife & Natural Environment (Scotland) Act 2011: Due Diligence Good Practice Guide'. <http://www.scottishlandandestates.co.uk/index.php?option=com_content&view=category&id=95&Itemid=128> (last accessed 20 December 2012).

Slee, B. (2009). 'Looking Backwards–Looking Forwards: Relocalisation and Sustainable Development'. Centre for Remote and Rural Studies Seminar Series 2008-9, UHI Millennium Institute, Inverness, 15 May 2009.

Stirling, A. (2006). 'Analysis, participation and power: justification and closure in participatory multi-criteria analysis'. *Land Use Policy* 23, pp. 95–107.

Swanwick, C. (2009). 'Landscape as an integrating framework for upland management'. In: Bonn, A., Allott, T., Hubacek, K. and Stewart, J. (eds) *Drivers of Environmental Change in Uplands*. Routledge, London and New York, pp. 339–57.

Tewdwr-Jones, M., Gallent, N. and Morphet, J. (2010). 'An Anatomy of Spatial Planning: Coming to Terms with the Spatial Element in UK Planning'. European Planning Studies 18 (2), pp. 239–57.

Thin, A. (2010). 'Address to Scottish Estates Business Group Half Year Meeting'. 22 April 2010, Edinburgh.

UK National Ecosystem Assessment (2011). 'The UK National Ecosystem Assessment: Synthesis of the Key Findings'. UNEP–WCMC, Cambridge.

Vasishth, A. (2008). 'A scale-hierarchic ecosystem approach to integrative ecological planning'. *Progress in Planning* 70, pp. 99–132.

Warren, C. R. (2009). *Managing Scotland's Environment*. Second Edition. Edinburgh University Press, Edinburgh. 490 pp.

Warren, C. R. and McKee, A. (2011). 'The Scottish Revolution? Evaluating the impacts of post-devolution land reform'. *Scottish Geographical Journal* 127 (1), pp. 17–39.

Wightman, A. (2010). *The Poor Had No Lawyers: Who owns Scotland and how they got it*. Birlinn, Edinburgh. 320 pp.

Index